Organic Farming

NIPA® GENX ELECTRONIC RESOURCES & SOLUTIONS P. LTD.
New Delhi-110 034

www.nipaers.com
Browse, Search, Read & Buy...
Print Books
eChapters
eBooks
Publishers
Forthcoming
Subject Catalogues
New Titles
Online Resources on:
✓ Current Affairs
✓ Reasoning, Logic & Aptitude
✓ Competitive Examinations
✓ English Language Lab
✓ Writing & Pronunciation Tools
✓ Personality Development
✓ Online Programmes for Professionals
✓ Interview Preparation
Pay Using
PhonePe
Google Pay
paytm
PayPal
UPI
CC Avenue

Organic Farming

A.K. Singh
Programme Coordinator/Senior Scientist (Soil)
ICAR Research Complex for NEH Region
Former Associate Professor (Agriculture)
North Eastern Regional Institute of Water and Land Management
(North Eastern Council)

NIPA® GENX ELECTRONIC RESOURCES & SOLUTIONS P. LTD.
New Delhi-110 034

NIPA® GENX ELECTRONIC
RESOURCES & SOLUTIONS P. LTD.

101,103, Vikas Surya Plaza, CU Block
L.S.C. Market, Pitam Pura, New Delhi-110 034
Ph : +91 11 27341616, 27341717, 27341718
E-mail: newindiapublishingagency@gmail.com
web: www.nipabooks.com

For customer assistance, please contact
Phone: + 91-11-27 34 17 17
Fax: + 91-11- 27 34 16 16
E-Mail: feedbacks@nipabooks.com

ISBN: 978-81-19235-99-5

Composed and Designed by NIPA®.

Preface

This book has been written to fill the needs of a general text material for conducting courses in Organic Farming for students studying in agricultural colleges and universities. The chapters systematically cover the principles and practices of organic farming. This book is divided into 7 chapters. Chapters cover basic concepts and relevance of organic farming; organic production requirement; biological intensive nutrient management; integrated pests, disease and weed management; quality control standards and certification; and organic market opportunities.

It will also be a standing guidance to the teachers for teaching organic farming in the new concept of agriculture education and development. I hope this book should provide good foundation in organic farming to students and academician of agriculture science and inspire them to take up further studies in the subject.

The authors duly acknowledge the sources of the diagrams and tables that have been reproduced from reports, journals, websites, books and other publications.

Dated: 14th January 2015 **A. K. Singh**

Udaipur

Contents

Acronyms

FAO	Food and Agriculture Organization
ITC	International Trade Center
IFOAM	International Federation of Organic Agriculture Movements
ICS	Internal Control Systems
LEISA	Low External Input and Sustainable Agriculture
NPOP	National Programme on Organic Agriculture
UNCTAD	United Nations Conference on Trade and Development
UNEP	United Nations Environment Programme
USDA	United States Department of Agriculture
FYM	Farmyard Manure
APEDA	Agricultural and Processed Food Products Export Development
NSOP	National Standard for Organic Production
JAS	Japan Organic Standard
MoC	Ministry of Commerce
SECO	State Secretariat for Economic Affairs
HACCP	Hazard Analysis Critical Control Points
OCCP	Organic Critical Control Points
GMO	Genetically Modified Organism
EU	European Unions
HYV	High Yielding Variety
WHO	World Health Organization
FAO	Food and Agriculture Organization
IP	Integrated Production
WTO	World Trade Organization
AIFOF	All India Federation of Organic Farming
NGO	Non-Governmental Organization
FIBL	Research Institute of Organic Agriculture
SOEL	Foundation of Ecology and Agriculture
IPM	Integrated Pest Management
VAM	Vesicular Arbuscular Mycorrhiza

Chapter - 1

Principles of Organic Farming

Farming is the intervention of man in the growth cycle of plants, animals or birds to get utility products of choice at maximum levels and of highest quality. The growing and caring of plants is called *Crop Husbandry*, and that of animals is called *Animal Husbandry*. Farming is our most basic activity. It is considered as the backbone of Indian economy. Farming not only nourishes the whole population, but also provides a means of livelihood to more than 60 per cent of our population. Our history, culture and community values are embedded in farming. Today when we think of farming, what comes to our mind are long stretches of farm land, farmer tilling the field, pumping water into fields, applying fertilizers or pesticides, plants with so many fruits, tractors with loads of farm produce, cross bred cows, a farm chicken, aquaculture, fish and so on. A more advanced image on this subject will be of the small plantlets in conical flasks or test tubes, what we call Tissue culture, High Yielding Varieties (HYV), Genetically Modified (GM) crops, Green House farming and other techniques that increase the crop yields. So all these activities are considered as the products of science that our scientists were able to develop over a period of time and helped the growing population to meet their food requirements.

1.1 The Science of Modern Farming

In independent India, more food production was needed to meet the growing food demands of the population. This increased demand for food had forced many countries including India to adopt various improved techniques and cultural practices like the use of high yielding variety (HYV) seeds, chemical fertilizers and pesticides to increase the farm production. This surge in production of foodgrains as a result of adopting HYVs, chemical fertilisers and pesticides is what we called the *"Green Revolution"*. The objective behind Green Revolution was the production of more food per unit of land within a short period of time. Chemicals in the form of fertilisers and pesticides play a major role in Green Revolution.

Today's chemical farms have little use for the skilled husbandry which was once the guiding principle of working the land. The emphasis today is solely on productivity - high input in exchange for high returns and productivity. Four important considerations - what happens to the land, the food it produces, the people who eat it and the communities which lose out - are overlooked.

The present modern farming includes-

Land exhaustion: The constant use of artificial fertilizer, together with a lack of crop rotation, reduces the soil's fertility year by year.

Fertilizers: High yield levels are produced by applying large quantities of artificial fertilizers, instead of by maintaining the natural fertility of the soil.

Nitrate run-off: About half of the nitrate in the artificial fertilizer used on crops is dissolved by rain. The dissolved nitrate runs off the fields to contaminate water courses.

Soil erosion: Where repeated deep ploughing is used to turn over the ground, heavy rains can carry away the topsoil and leave the ground useless for cultivation.

Soil compaction: Damage to the structure of soil by compression is a serious problem in areas that are intensively farmed. Conventional tillage may involve a tractor passing over the land six or seven times, and the wheeling can cover up to 90 per cent of a field. Even a single tractor pass can compress the surface enough to reduce the porosity of the soil by 70 per cent, increasing surface run-off and, therefore, water erosion. In the worst cases, the surface run-off may approach 100 percent - none of the water penetrates the surface.

Biocide sprays: Pest and disease incidence in plants has become a major hindrance/bottle neck for crop production. Use of HYVs along with chemical fertilisers has increased the menace by making plants more succulent and thereby attracting more pests and disease causing pathogens. Toxic chemical pesticides and insecticides are used for controlling pests and diseases. Diseases are mainly caused by plant pathogenic fungus, bacteria or virus; these are also controlled by chemical fungicides, antibiotics, bactericides etc. Most of these chemical pesticides are non-biodegradable and are highly toxic in nature. Most crops receive many doses of different chemicals before they are harvested.

Cruelty to animals: On most 'modern' farms, all animals are crowded together indoors. Complex systems of machinery are needed to feed them, while constant medication is needed to prevent disease. The cruelty involved in managing, breeding, growing and slaughtering farm animals today is unimaginably repulsive and horrifying.

Animal slurry: With so many animals packed together in indoor pens, their manure accumulates at great speed. It is often poured into ponds which leak into local watercourses, contaminating them with disease-causing organisms and contributing to algae-blooms.

Imported animal feed: Many farms are not self-sufficient in animal feed; instead they rely on feed brought into the farm.

Stubble burning: Stubble is burned, large amounts of potentially useful organic matter disappear into the sky in clouds of polluting smoke.

Loss of cultivated biodiversity: Large and other chemical farms tend to be monocultures growing the same crop and crop variety.

Threat to indigenous seeds and animal breeds and species: Native cultivars and animal breeds lose out to exotic species and hybrids. Many native animal breeds are today threatened with extinction. The same holds true for many indigenous plant varieties which have disappeared within the space of one generation.

Habitat destruction: Agribusiness farming demands that anything which stands in the way of crop production is uprooted and destroyed. The wild animals and plants which were once a common sight around farms are deprived of their natural habitat and die out.

Contaminated food: Food, both plant and animal products, leaves the farm contaminated with the chemicals that were used to produce it.

Destruction of traditional knowledge systems and traditions: Rural indigenous knowledge and traditions, both agricultural and non-agricultural, is invariably connected to agriculture and agricultural systems.

Control of agriculture inputs and food distribution channel: The supply and trading in agricultural inputs and produce is in the hands of a few large corporations. This threatens food security, reducing the leverage and importance of the first and the last part of the supply chain - the farmer and the consumer.

Threat to individual farmers: Chemical agriculture is a threat to their livelihoods and changes their lifestyles, unfortunately not for the better.

Prolonged and over dose of these chemicals are hazardous to human health, soil and environment. The food we produce now contains the residues of toxic chemicals that are harmful to our body. The air that we breath is not fresh, as it is highly polluted with harmful chemicals. There is also unusual phenomenon like acid rain. The excess chemical from the field get washed off by rain, reaches water bodies and affects aquatic life. Water that is available for drinking purpose also contains traces of that chemicals and they can damage health and internal organs. It is estimated that over 20,000 people die in India every year due to exposure to toxic chemicals. Many new diseases are reported in humans, animals and plants. The scientists are unable to resolve them.

Ecological imbalance and environmental pollution have adversely affected farming and farm products. This calls for corrective farming practices that can produce sufficient quantity of food without harming the environment. Here comes the relevance of the ***Organic Farming*** which is a traditional, viable, low input use and sustainable method of agriculture to produce healthy food. It also helps to reduce environmental pollution and can strengthen the natural resource base. It ensures sustainable farm production.

1.2 Development of Organic Farming

Throughout the 1950s, and 60s, the main thrust of farming was on increasing the crop and animal productivity so as to satisfy the immediate needs of the growing population and to attain food self-sufficiency. Organic Farming was not viewed very favourably by the scientists or farmers. By the end of the 1980s and especially in the 1990s, organic farming came to the forefront in response to the growing awareness of environmental conservation issues.

Milestones in organic farming movement

1924	Demeter biodynamic label founded
1940	Sir Albert Howard publishes ***An Agricultural Testament***
1942	J.I. Rodale publishes ***Organic Farming and Gardening***
1943	Lady Eve Balfour publishes ***The Living Soil***
1946	Soil Association founded in the UK
1967	Soil Association publishes first Organic Standards
1972	Founding of IFOAM
1974	Oregon State (US) adopts Organic Legislation
1979	First California Organic Foods Act adopted
1980	IFOAM Basic Standards published
1985	France adopts organic legislation
1990	Organic Foods Production Act passed in the United States
1991	EU Regulation 2092/91 adopted
1992	Establishment of the IFOAM Accreditation Programme
1999	Codex Alimentarius guidelines adopted
1999	EU organic livestock regulation published
2000	Japanese organic regulation published
2002	Indian Organic Regulation (NPOP -National Programme for Organic Production) published.

Many new organisations were formed by producers, consumers and other groups interested in an eco-friendly lifestyle more in tune with nature. These organisations draw up their own specifications governing crop production, right from sowing to the final stage of reaching the consumers. It was in the 1990s that organic farming really took off when the production grew up along with the consumer interests in its products all over the world. Today farmers in the developed and developing countries are encouraged to convert their farms to ***Organic Farms.*** Organic Farming is one of the fastest growing sectors of agricultural production and plays an important role as an alternative and safe food production process. This system assures the production of high quality food while conserving the soil fertility and enhancing its fertility.

1.2.1 Role of international organisations

The formation of the International Federation of Organic Agriculture Movements (IFOAM) in 1972 gave an international framework to the discussion and codification of collective principles and procedures of organic farming. It brings together organisations from all over the world that are involved in organic production, certification, research, education and promotion. Its specifications and standards are indicative and act as guidelines to the state-of-the-art methods of production and processing of organic products. FAO (Food and Agriculture Organisation) and WHO (World Health Organisation) have declared that international guidelines on organically produced food products should be considered important for consumer protection and trade facilitation.

The Codex Alimentarius Commission, an international food standards body jointly formed by FAO and WHO, has developed guidelines for the production, processing, labelling and marketing of organically produced food. The Codex Guidelines are important for equivalent judgment under the rules of World Trade Organization (WTO).

1.3 What is Organic Farming?

The word *'organic',* means 'living', 'earth friendly' or 'of plant or animal origin'. It contrasts with the farming wherein chemical inputs are used. It also denotes the produces that have been produced in accordance with certain norms and standards during cultivation, handling, processing and marketing stages. It is certified by a certifying body.

Sometimes the term 'organic agriculture' is used in place of 'organic farming'. There is no difference in the meaning of these terms. In general, people consider that it is the kind of agriculture that is based on application of organic manures, natural inputs, and pesticides of plant origin. For others, it denotes the agricultural systems that follow the principles and logic of a living organism in which all elements (soil, water, air, plants, animals, farmer etc.) are closely linked with each other.

Organic Farming includes all agricultural system which promotes production practices that are environmentally, socially and economically sound. It considers the local soil health as a key to the successful production. By respecting the natural capacity of plants, animals and the landscape, it aims to optimize quality in all aspects of agriculture and the environment. Thus, Organic Farming helps in attaining an eco-friendly and sustainable farm production.

1.3.1 Definition of organic farming

The coinage of the term ***'organic farming'*** is credited to ***Lord Northbourne.*** The term is derived from the concept of "the farm as (living) organism" which he describes in his book **'Look to Land'** (1940). He described organic fanning as a holistic and ecologically balanced approach of agriculture.

. .. the term 'organic' is best thought of as referring not to the type of inputs used, but to the concept of the farm as an organism, in which all the components - the soil minerals, organic matter, micro-organisms, insects, plants, animals and humans - interact to create a coherent, self-regulating and stable whole.

To define the concept of organic farming, we may refer to the definition developed by the **Codex Alimentarius**, on the basis of inputs received from experts from different countries.

According to the Codex, "organic agriculture is a holistic production management system which promotes and enhances agro ecosystem health, including biodiversity, biological cycles and soil biological activity". It emphasizes the use of management practices in preference to the use of on-farm inputs taking into account that regional conditions require locally adapted systems. This can be accomplished by adopting cultural, biological and mechanical methods wherever possible as against the use of synthetic materials.

According to the International Federation of Organic Agriculture Movements (IFOAM), "organic agriculture is an agricultural system that promotes environmentally, socially and economically sound production of food, fiber, timber, etc. In this system, soil fertility is seen as the key to successful production. Working with the natural properties of plants, animals and the landscape, organic farmers aim to optimise quality in all aspects of agriculture and environment".

The United States Department of Agriculture (USDA) has defined organic agriculture as "an ecological production system that promotes and enhances biodiversity, biological cycles and soil biological activity. It is based on minimal use of off-farm inputs and relies on farm management practices that restore, maintain and enhance ecological harmony".

Conceptually all these aspects and dimensions are integrated harmoniously in organic farming and can be operationally defined as "Organic farming is the systematically and functionally integrated and balanced farming system facilitating best yield expressions and substantial automatic regeneration".

1.4 Concept of Organic Farming

The concept of organic farming is based on related aspects like 'organic concept', 'holistic concept', 'living soil concept' and 'healthy plant concept' where nature is perceived to be more than just an individual element. In this farming system there is dynamic interaction between soil, plant, human beings, animals, eco-system and environment. It is more of a process based farming approach where the quality of produce is given importance. Farmers are encouraged to convert their existing farms into organic farms. A farmer willing to go in for organic farming should further understand the broad areas covering these concepts.

1.4.1 Organic concept

The term 'organic' is not directly related to the type of inputs (fertilizers, HYV seeds, labour etc) used, but refers to the concept of the farm as a living organism, in which all the supporting factors such as the soil, microorganism, living creatures, plants, animals and human beings - interact to create a healthy environment for the plants and animals to grow and produce healthy food for the human being. Every operation a farmer does in his farm is evaluated against ecological, economical and social imperatives with an awareness of local and global implications.

1.4.2 Holistic concept

The 'holistic concept' is to view the agro-ecosystem as a whole in relation to the farm. Farmers are aware that their action in one sphere can affect other parts of the ecosystem for better or worse. They take advantage of the biological cycles and interaction between animals, plants, soil and climate. These are fundamental to enhance the ecosystem's productivity, health and efficiency. Basically organic farmers attempts to work with natural cycles rather than disrupting them.

1.4.3 Living soil concept

The productive capacity of soil is linked with its health, which in turn is influenced by the 'humus' or 'organic matter content'. The 'humus' is the most dynamic component of a **living soil.** It plays an important role in shaping up the soil's physical, chemical and biological properties. Regular replenishment of 'humus' after harvest of a crop and its maintenance at optional levels are of primary importance into sustaining the productive function of the 'living soil'. Apart from being a store house of many plant nutrients, the 'humus' improves the soil structure, activates soil life, minimizes the leaching loss of nutrients and strengthens natural resistance of plants against pest and diseases.

On the other hand, chemical fertilizers in the absence of 'humus' or inadequate levels of soil organic matter, will impair soil structure, disturb soil biotic, cause unfavourable changes in soil reaction and weaken the natural resistance of plants against adverse agro-climatic conditions and pest and disease attack.

1.4.4 Healthy plant concept

In modern farming, the high yield varieties have displaced the diverse and locally adapted indigenous cultivars. As these high yielding crop varieties are with narrow genetic base, they are more susceptible to pest and disease attacks compared to the local indigenous varieties. This has resulted in a spurt in the incidence of pests and diseases, necessitating increased use of pesticides. Prolonged use of the pesticides against a particular insect helps it in developing resistance to these chemicals. It also kills many of the naturally occurring predators and parasites, which arc biological pest control agents. In many cases, pests that caused minor damages in the past have become major threat in the absence of natural enemies. This is termed as **Pest Resurgence.**

Similarly, in plant diseases too, spraying of fungicides had resulted in building up of resistance by harmful pathogens and micro organisms. The build up of higher levels of these toxic chemical residues in the plant, food, atmosphere, soil, ground water and other water bodies is the most serious health and environmental problems associated with the continued use of chemicals.

Thus, the organic farming is not mere growing of plants and animals but it needs a thorough understanding of various principles and prudent management of on-farm resources and process.

1.5 Principles in Organic Farming

The Principles of Organic farming reflect the potential contribution this farming system could makes to the world, and holds out a vision of how all agricultural system should function in the larger context. The four principles as delineated by IFOAM are:

- The Principle of Health
- The Principle of Ecology
- The Principle of Fairness
- The Principle of Care

Each principle is articulated through a statement followed by an explanation. These principles are to be used as a whole. They are composed as ethical principles to inspire action.

Principle of health: *Organic Agriculture should sustain and enhance the health of soil, plant, animal, human and planet as one and indivisible.*

This principle points out that the health of individuals and communities cannot be separated from the health of ecosystems - healthy soils produce healthy crops that foster the health of animals and people.

Health is the wholeness and integrity of living systems. It is not simply the absence of illness, but the maintenance of physical, mental, social and ecological well-being. **Immunity, resilience** and **regeneration** are key characteristics of **health**.

The role of organic agriculture, whether in farming, processing, distribution, or consumption, is to sustain and enhance the health of ecosystems and organisms from the smallest in the soil to human beings. In particular, organic agriculture is intended to produce high quality, nutritious food that contributes to preventive health care and well-being. In view of this it should avoid the use of fertilizers, pesticides, animal drugs and food additives that may have adverse health effects.

Principle of ecology: *Organic Agriculture should be based on living ecological systems and cycles, work with them, emulate them and help sustain them.*

This principle roots organic agriculture within living ecological systems. It states that production is to be based on ecological processes, and recycling. Nourishment and well-being are achieved through the ecology of the specific production environment. For example, in the case of crops this is the living soil; for animals it is the farm ecosystem; for fish and marine organisms, the aquatic environment.

Organic farming, pastoral and wild harvest systems should fit the cycles and ecological balances in nature. These cycles are universal but their operation is site-specific. Organic management must be adapted to local conditions, ecology, culture and scale. Inputs should be reduced by reuse, recycling and efficient management of materials and energy in order to maintain and improve environmental quality and conserve resources.

Organic agriculture should attain ecological balance through the design of farming systems, establishment of habitats and maintenance of genetic and agricultural diversity. Those who produce, process, trade, or consume organic products should protect and benefit the common environment including landscapes, climate, habitats, biodiversity, air and water.

Principle of fairness: *Organic Agriculture should build on relationships that ensure fairness with regard to the common environment and life opportunities.*

Fairness is characterized by equity, respect, justice and stewardship of the shared world, both among people and in their relations to other living beings.

This principle emphasizes that those involved in organic agriculture should conduct human relationships in a manner that ensures fairness at all levels and to all parties - farmers, workers, processors, distributors, traders and consumers. Organic agriculture should provide everyone involved with a good quality of life, and contribute to food sovereignty and reduction of poverty. It aims to produce a sufficient supply of good quality food and other products. This principle insists that animals should be provided with the conditions and opportunities of life that accord with their physiology, natural behaviour and well-being.

Natural and environmental resources that are used for production and consumption should be managed in a way that is socially and ecologically justified and should be held in trust for future generations. Fairness requires systems of production, distribution and trade that are open and equitable and account for real environmental and social costs.

Principle of care: *Organic Agriculture should be managed in a precautionary and responsible manner to protect the health and well-being of current and future generations and the environment.*

Organic agriculture is a living and dynamic system that responds to internal and external demands and conditions. Practitioners of organic agriculture can enhance efficiency and increase productivity, but this should not be at the risk of jeopardizing health and well-being. Consequently, new technologies need to be assessed and existing methods reviewed. Given the incomplete understanding of ecosystems and agriculture, care must be taken.

This principle states that precaution and responsibility are the key concerns in management, development and technology choices in organic agriculture. Science is necessary to ensure that organic agriculture is healthy, safe and ecologically sound. However, scientific knowledge alone is not sufficient. Practical experience, accumulated wisdom and traditional and indigenous knowledge offer valid solutions, tested by time. Organic agriculture should prevent significant risks by adopting appropriate technologies and rejecting unpredictable ones, such as genetic engineering. Decisions should reflect the values and needs of all who might be affected, through transparent and participatory processes.

The principles as per International Federation of Organic Agriculture Movement (IFOAM):

- To produce sufficient quantities of high quality food, fibre and other products.

- To work compatibly with natural cycles and living systems through the soil, plants and animals in the entire production system.
- To recognize the wider social and ecological impact of land within the organic production and processing system.
- To maintain and increase long-term fertility and biological activity of soils using locally adapted cultural, biological and mechanical methods as opposed to reliance on inputs.
- To maintain and encourage agricultural and natural biodiversity on the farm and surrounds through the use of sustainable production systems and the protection of plant and wildlife habitats.
- To maintain and conserve genetic diversity through attention to on-farm management of genetic resources.
- To promote the responsible use and conservation of water and all life therein.
- To use, as far as possible, renewable resources in production and processing systems and avoid pollution and waste.
- To foster local and regional production and distribution.
- To create a harmonious balance between crop production and animal husbandry.
- To provide living conditions that allows animals to express the basic aspects of their innate behaviour.
- To utilise biodegradable, recyclable and recycled packaging materials.
- To provide everyone involved in organic farming and processing with a quality of life that satisfies their basic needs, within a safe, secure and healthy working environment.
- To support the establishment of an entire production, processing and distribution chain which is both socially and ecologically responsible.
- To recognize the importance of, and protect and learn from, indigenous knowledge and traditional farming systems.

1.6 Characteristics of Organic Farming

One of the notions that we have about farming is to produce large quantities of food materials from the available land. In case of organic farming, there are several compelling principles that characterize the certified organic farm. They include-

- Biodiversity
- Diversification and integration of enterprises
- Sustainability
- Natural plant nutrition
- Natural pest management, and
- Integrity.

If we examine some of the successful organic farms around us, most of above operations are reflected in the activities of these farms to a greater or lesser degree. Since each farm is a distinct entity, there is a large degree of variation also. Now let us study these one by one.

Biodiversity: As a general rule, diverse ecosystems in nature (forest) have a higher degree of stability than those with only a few species. This is true for agro-ecosystems also. Farms with a diverse mix of crops have better chance of supporting beneficial organisms that assist in pollination and pest management. Diversity above ground also suggests diversity in the soil, providing better nutrient cycling, disease suppression, tilth, and nitrogen fixation. Good organic farmers mimic the biodiversity of nature through practices like intercropping, companion planting, establishment of beneficial habitats, and crop rotation. The effort to increase biodiversity works hand-in-hand with enterprise diversity, which is often an objective on organic farms.

Diversification and integration of enterprises: The drive to build biodiversity in organic systems encourages diversity among enterprises, but not as isolated or independent entities. Good organic operations integrate their various enterprises, especially crops and animals. A good example is the farmers of Kerala state who maintain organic crop and livestock operations.

A typical organic farmer integrates all the farm operation in such a way that he ties the needs of crops and livestock together in a practical and elegant way. The forage and grain needs of ruminant livestock are made from a diverse mixture of fodder crops. Forage legumes in rotation fix a sustainable supply of nitrogen in the soil that feeds subsequent non-legume crops in rotation. Manure from the livestock enterprises is converted through biogas plants and vermicomposting as a nutrient resource and recycled back to the crop fields.

Farms such as these have, the additional advantage of greater economic sustainability, as their risks are spread over livestock, crop enterprises and value addition of produce.

Sustainability: In addition to the greater economic sustainability afforded by enterprise diversification, organic farmers are often able to reap market premiums for certified production. However, since many organic enterprises realize somewhat lower marketable yields, this has not always translated into higher profits or greater economic sustainability. As more and more organic growers enter the marketplace, it is likely that premiums will stabilize at modest levels and may vanish for some crops. Organic producers need to look well ahead and be aware of shifting trends and seasonality of crops.

Organic farms perform well on many of the measurable indicators associated with sustainability, such as energy consumption and environmental protection. However, sustainability is an ideal, and the best that can be said is that many of the organic farms are closer to the ideal than most alternatives – certainly closer than comparable chemical farming operations.

Natural plant nutrition: Plants photosynthesize and make sugars, which are ultimately synthesized into proteins and other plant yields. Human and animals, by contrast, can obtain energy, foods, proteins, and vitamins only by consuming plants or animals products/byproducts. Both plants and animals also require minerals. Humans and animals extract minerals, along with sugars and proteins from the food they eat. Plants, too, obtain minerals - and a wide range of vitamins, antibiotics, and other useful compounds - through digestion. However, plant digestive systems are not internalized as they are in animals. Plants must rely on the external digestive processes of the soil system within reach of their roots – a zone called the *rhizosphere*.

The organic philosophy of crop nutrition begins with proper care and nourishment of the organisms responsible for the soil digestive process. Organic farmers believe this is best accomplished by avoiding toxic chemicals and practices - like excessive tillage - that are harmful to soil organisms, as well as by the addition of organic matter and natural rock minerals. Conventional systems, in contrast, try to circumvent the soil's digestive process and provide needed minerals to the plant directly in a soluble form.

From the organic perspective, the chemical farming approach has several flaws.

- Applying large quantities of soluble fertilizer like urea, ammonium sulphate or super phosphate to a crop a couple of times per season could cause nutritional imbalances that lead to crop diseases, insect infestations, and reduced food quality.
- Failure to support and care for soil biotic life, along with other practices that are down right destructive, ultimately leads to its decline. As a result, plants lose out on the vitamins and other beneficial products these organisms produce, tilth is reduced, and the soil becomes increasingly dependent on synthetic inputs.
- Chemical fertilization tends to concentrate on a limited number of

macronutrients, even though the need for at least 16 soil minerals is scientifically recognized. This skewed focus is also responsible for generating imbalances in the plant.

- Application of large amounts of soluble nutrients can stimulate certain problem weed species.
- Soluble nutrients - especially nitrate - are prone to leaching, which can cause a number of environmental and health problems. It is organic farming's approach to soil building and plant fertilization that is the true basis for the belief that organic food and feed have superior nutritional value.

Natural Pest Management: Whether it is chemical or organic farming, all farmers are concerned with pests. They spend a lot of time and resources in controlling them. However, in the organic farms, the pests - whether weeds, insects or diseases - are not simply scourges. They are indicators of how far a production system has strayed from the natural ecosystems it should imitate. Certain weeds, for example, tend to predominate when soils are too acidic or too alkaline; some become a problem when soil structure is poor and conditions become anaerobic; others may be stimulated by excessive fertilizer or manure salts.

Organic proponents also believe that insect pests are attracted to inferior or weak plants - the result of poor crop nutrition. Their logic continues by asserting that pests are naturally repelled by vigorous and well-nourished plants. This belief is often challenged, and significant research remains to be done. As scientific understanding has grown, insect pest outbreaks are also being understood as imbalances in the whole agro ecosystem and how it is managed.

In nature, massive pest outbreaks are relatively rare and short-lived. Due to the presence of natural predators, parasites, and disease agents that quickly knock the pest numbers back down to a moderate level. In farming systems that inadvertently destroy or otherwise fail to support the natural control complex, pest problems are routine and deteriorate with time. The farmer becomes increasingly addicted to costly and extreme control methods to produce a crop. Most organic growers consider pesticides to be a cause of agro-ecosystem imbalances and employ allowed natural pesticides as little as possible.

Integrity: Integrity refers to the systems in place and actions undertaken to assure that consumers of organic products get what they pay for. Consumers have a right to expect that the organic food they buy not only be raised by organic methods but be protected from contamination and from non-organic products.

While the responsibility for much of this rests with others in the organic marketing chain, many certified organic growers need to incorporate additional practices that work to assure the integrity of their products. Proper record keeping is very important in this regard, though growers are often reluctant to spend much time on it. Among the more important production practices in the field are buffer strips, which reduce chemical drift from neighbouring fields and roadsides, while also serving water and soil conservation objectives.

1.7 Distinction from other Farming Systems

Traditional farming: Traditional farming in India dates back to the Neolithic age of 7,500-6,500 BC. The farmers of ancient India are known to have evolved nature friendly farming practices. They considered the Earth as a living being. Agriculture was not developed just as a production system, but as a culture. Great attention was paid to agricultural technologies and agronomic practices. Sophistication came in through genetic diversity, crop rotation and mixed cropping systems.

Animal husbandry was an integral part of the farming practice. Classical Indian plant science, *Vrikshayurveda* (Science of Plants) and *Mrgayuveda* (Science of Animals) in the form of Sanskrit hymns are a corpus of rich textual knowledge. In Vrikshayurveda, practices related to collection, selection and storage of seeds, germination, sowing, various techniques of plant propagation, grafting, nursing and irrigation, testing and classification of soil, selection of soil suitable to various plants types of plants, manuring, pest and disease management, preventive and promotive care to build up disease resistance and to cultivate healthy plants are detailed.

Favourable and unfavourable meteorological conditions were taken care of. Plants were used as indicators of weather, water, minerals etc. This knowledge system is alive even today among millions of Indian farmers. Furthermore, it is kept alive in many forms, such as folk songs, rituals, proverbs and riddles. Traditional farming practices are very rarely seen among the modern farmers today. However, some farmers in a few farming communities still follow these practices.

Today organic farmers make use of these indigenous knowledge systems and try to integrate them in the modern organic agricultural practices, thus making changes easier and more effective, *eg.,* use of Panchagavya growth stimulant/ pest control formula and Njattuvela crop calendar followed in Kerala state of India.

Many methods and techniques of organic agriculture have originated from various traditional farming systems all over the world. Whether a certain traditional farming system can be called organic will depend on whether all the organic standards are fulfilled. For instance, some traditional systems are in conflict with the requirements of organic animal husbandry (e.g. sufficient space and free movement), the necessary prevention of soil erosion, the ban on cutting forests and burning biomass (e.g. slash and burn systems).

Bio-dynamic farming: This is a holistic system of agriculture that seeks to connect nature with cosmic creative forces. Bio-dynamic farming method has developed since 1922, on a foundation of advice and instruction given by Rudolf Steiner, a German Philosopher, known for his world view called Anthroposophy (Wisdom of man). The name 'Bio-dynamic' refers to cosmic energies, that create and maintain life. This name was adopted by the first group of farmers inspired by Rudolf Steiner's advice. The term was derived from two Greek words 'bios' (life) and 'dynamics' (energy). The use of the word 'method' indicates that one is not dealing merely with the production of another fertilizer, which is organic, but rather, involves certain principles. These principles on practical application can provide healthy soil and healthy plants that in turn produces healthy food for man and animals. This approach has now become a distinct farming strategy of organic farming and the produces are certified as 'DEMETER'.

Natural farming: Natural farming, also known as *do-nothing farming* or *no-till farming* was popularised by Masanobu Fukuoka, a Japanese agriculturist in 1940. He outlined his experience in his famous compilation, '*The One-straw Revolution*'. The most essential aspect of natural farming is to let nature play a dominant role to the maximum extent possible. No-tillage, farm biodiversity, symbiotic farm inputs and protection of soil cover are given importance in this method of farming. The *seed ball technique* for sowing has also been highlighted by Fukuoka. The physical work and labour inputs are 80% less compared to other farming systems. Fukuoka proved that he can achieve yields similar to those of chemical farming. His methods have also been adapted to suit European conditions and many farmers practice it there. In India, Fukuoka is fondly regarded and his work has found a number of practitioners who have termed their method of farming as 'rishi kheti' literally meaning '*agriculture of the sages*'.

The essence of Fukuoka's method is to reproduce natural conditions as closely as possible. There is no ploughing, as the seed germinates quite happily on the surface if other conditions are favourable. There is also considerable emphasis on maintaining diversity. A ground cover of white clover grows under the grain plants to provide nitrogen. Weeds are also considered part of the ecosystem, periodically cut and spread on the surface so the nutrients they contain are returned to the soil. Ducks are let into the grain field, and specific insectivorous carp are also introduced into the rice field at certain times/seasons of the year to eat slugs and other pests.

The ground is always covered with clover and weeds. The straw from the previous crop is used as mulch. Each grain crop is sown before the harvest of the standing crop. This is done by broadcasting the seed among the standing crop. He also re-introduced the ancient technique of *seed balls* (Earth dumpling). The seed for next season's crop is mixed with clay, compost, and sometimes manure, and formed into small balls. The result is a denser crop of smaller hut highly productive and stronger plants.

Sustainable agriculture: It is estimated that about 65 percent of India's cropped area is rainfed. It is mostly seen in the dryland tracts. One speciality of this region is that high-input demanding crops are not grown on these lands. The fertiliser and pesticide use on these drylands is always much less. This is because the chemical fertilisers require sufficient water to respond. Also the economics of these hardy or 'not so profitable' crops will not permit expensive inputs. These areas though not "organic", can be considered as 'sustainable farming' as the input use is less. However, produce from these areas cannot be sold as organic, as the certification agency will object to the application of certain chemicals even in smaller quantities. But farmers can opt for converting the farm to organic over a period by following certain guidelines. These approaches exclusively rely on use of on farm resources. As far as goals are concerned, sustainable agriculture therefore has much in common with organic agriculture. Therefore, systems which do use chemical fertilizers, pesticides or genetically modified organisms are classified as sustainable.

LEISA farming approach: A more recent approach in agriculture, propagated by ILEISA (Information Centre for Low-External Input and Sustainable Agriculture) of ETC Foundation, the Netherlands, is the LEISA (Low-External Input and Sustainable Agriculture) farming. It states that when 'modern' external inputs are not available or profitable, farming has to depend on optimal use of local resources, human and animal labour, ecological process, recycling and site-specific genetic resources.

Production levels have to be raised, but at the same time it is necessary to stop depletion and degradation of the resource base (soil, nutrients, vegetative cover, genetic resources, indigenous knowledge and social cohesion). Self-reliance, local economies, strong local institutions and local 'agri-culture' are important to prevent loss of sustainability. Intensification and increase of efficiency is possible to a certain extent by natural nitrogen fixation, mobilization and concentration of nutrients, diversification and better integration of different elements and activities. However, an important precondition is that the nutrient flows are kept in balance.

When natural external inputs such as sediments and biomass from outside the farm are scarce and population is growing fast, 'modern' external inputs may be needed. It would compensate for unavoidable losses of nutrients, invest in soil fertility, repair degradation, and bring down severe pest attacks. However, they should be used prudently and strategically, and are complementary to natural, external and internal inputs.

Permaculture: Permaculture, the word coined in Australia during late 1970 by Australian ecologist Bill Mollison, is gaining importance now. Permaculture means 'Permanent agriculture' defined as a system of farming that holds the prospects of sustained rather than temporary yields from the land. Mollison (1989) states, "*Sustainable use of land cannot be separated from personal lifestyle and philosophical matter*". It encourages a style of farming that puts pressure on the land for short-term gain which is detrimental to its long-term ability to produce food.

Permaculture, basically means a system of agriculture that relies on perennial crops, like fruits and nut trees, because of their ability to provide a more permanent and sustained harvest. It stresses the careful use of resources, agro-forestry, self-reliance and small-scale operations carried out as near as possible to the food consumer. Mollison suggested that 'earth repair work' was needed and it would yield a good economic return. He advocated that Polyculture (multiple species) forestry, small organic farms (in urban as well as rural areas) and recycling could facilitate high rates of return.

Integrated Production (IP): Integrated Production (IP) has gained importance over the last few years, especially in economies of transition and in industrialized countries. It does not refrain from using agro-chemicals, but aims at a reduction of its application. For plant protection, a combination of bio-control methods and chemical pesticides is used (Integrated Pest Management). If damage by pest or disease reaches defined threshold levels, chemical pesticides are applied. For plant nutrition, chemical fertilizers can be used, but usually maximum amounts are defined. Herbicides also are used. The regulations on IP are not always very clear and vary from country to country, if formulated at all. A few countries have developed labels and a control system for integrated production. In some countries integrated systems are called 'green production'. Above all, integrated production follows the same approach as conventional agriculture, and is far from the holistic understanding of organic agriculture. However, it can contribute to a healthier environment as it is easier for a large number of farmers to follow.

Conventional Vs Organic farming: Organic and conventional agriculture belonged to two different paradigms. The fundamental difference between the two competing agricultural paradigms are as follows:

Conventional Farming	Organic Farming
Centralization	Decentralization
Dependence	Independence
Competition	Community
Domination of nature	Harmony with nature
Specialisation	Diversity
Exploitation	Restraint

In contrast, several agro-ecologically based researchers stress more the fluid transition between conventional, integrated and organic farming, as an outcome of different assessments of economic, ecological and social goals. Consequently, technique strategies such as integrated pest management of balanced nutrient supply might improve conventional agriculture to such as an extent that it may appear unnecessary to strictly ban pesticides and mineral fertilizers as required by organic standards.

However, there is scientific reason that organic agriculture differs from conventional agriculture not only gradually but fundamentally. Implementing organic methods consequently seems to provide a new quality in how the agro-ecosystem works. This functioning cannot be explained by summing up single ecological measures. Organic farming seems to improve soil fertility in a way and to an extent which cannot be achieved by conventional farming even if the later consistently respects some ecologically principles.

Organic agriculture is one of several to sustainable agriculture and many of the techniques used (e.g. inter-cropping, rotation of crops, double digging, mulching, integration of crops and livestock) are practiced under various agricultural systems. What makes organic agriculture unique, as regulated under various laws and certification programmes, is that:

1) Almost all synthetic inputs are prohibited, and

2) Soil building crop rotations are mandated.

The basic rules of organic production are that natural inputs are approved and synthetic inputs are prohibited, but there are exceptions in both cases.

1.8 Why Do People Choose Organic Food?

Consumers of organic products continue to buy these produce from known sources. A survey among these consumers as well as proponents of organic farming revealed the following:

Better taste: Organic food tastes better; this is the simple message given by many people who eat organic.

Healthier: Organic foods have far less residues of pesticides, growth promoters and antibiotics. It has been shown in a number of studies that organic food contains more vitamins, nutrients and cancer-fighting antioxidants than non-organic food.

Animal friendly: Organic farming places great emphasis on animal welfare and is more animal friendly, where animals are not treated with synthetic growth hormones or drugs.

GMO free: Genetically Modified Organisms are not allowed in organic agriculture.

Fair play: By avoiding expensive inputs of agrochemicals, organic farmers can manage to reduce production costs and get out of the debt trap. Fair prices additionally can enable farmers to get reasonable returns for their efforts.

More sustainable: Organic production aims, at all times towards sustainability while having best management of the environment and wildlife as a priority. This implies a sustainable management of natural resources; soil water and bio diversity. Organic systems are less dependent on non-renewable resources.

Traditions meeting science: Organic systems rely on a modern and scientific understanding of ecology and soil science, while also integrating traditional agricultural knowledge.

1.9 Facts and Fallacies of Organic Farming

In spite of the massive movement towards organic farming philosophy and practices, there are a number of fallacies that surround organic farming, both in favour of and against. Some of the more common ones are:

Yields in organic farming are lower than chemical farming: When principles of organic farming properly followed, yields in organic farming are, in the long run, far greater than those obtained by chemical farming. In horticulture crops, the effects are even better. A look at agriculture in Punjab will tell the full story. The origin of the 'Green Revolution', Punjab's agricultural yields have remained the same for the past many years while the quantity of agro-chemicals required to maintain these static yields have steadily increased.

In the case of a chemical farm converting to organic however, there is often a loss in yield in the initial years and it takes a few years before yields increase and stabilise at a level often higher than that achieved under a chemical regime. It is therefore recommended to convert a farm to organic gradually over a period of three to four years if income from the farm is a key issue.

Organic farming is not economical: While certain practices such as composting and mulching do entail greater costs on account of labour, the overall cost of cultivation is usually lower than chemical farming. An important point to note here is that the farmer has to be self-sufficient in his requirement especially for the inputs like composts and pest control measures, otherwise economics do get skewed. Farmers who could make their own inputs are successful organic farmers.

Problem of supply of enough nutrients by using composts: This is a common argument put forth against organic farming, especially academicians. If one calculates the percentages of nitrogen, potassium and phosphorous in fertilisers and composts, the difference is indeed vast. Going by these 'scientific' calculations, one may find that in lieu of say 200 kgs of mixed fertilisers, one needs over 30 tons of composts, fairly impossible to supply two or three times a year. In organic farming however, the concept of feeding the plant does not exist. The attempt here is to feed the soil, keep it healthy and living and keep a PROCESS in motion. Much of the work is done by the numerous soil organisms and microorganisms that thrive in 'living' soils. The various practices of organic farming ensure that soil fertility is maintained and this symbiotic relationship is kept alive and vibrant through out the year rather than for a short period. The analogy we can think of is meeting your day's entire nutritional requirements (calculated by a dietician) by swallowing a few tablets and capsules. Your nutritional needs are met, but will you remain healthy?

There is big money in organic farming: So far in India, most organic farmers have turned (or in some cases remained) organic because of this belief. The 'organic market' exists for a small number of farmers who have access to a few specialised outlets. Otherwise, till consumers preference changes, it is difficult to see every organic farmer getting a premium for his undoubtedly superior product. An option for exports has also opened in the last decade which organic farmers can explore only if they are 'certified as organic'. There are a few internationally-recognised local agencies operating from India who inspect and certify organic farms. But because of certification cost and procedural huddles farmers are shying away from this.

The ICS (Internal Control System) mode of certification can be worked out to make certification process more farmer friendly and cost effective. Thus the organic farmers get fair price (not big money) when compared to their counterparts. Now attempts are on to bring organic growers and interested organic buyers together at least once in a week.

The non-use of chemical fertilisers and pesticides is organic farming: It may be mistaken that simply avoiding the use of chemical fertilisers and pesticides is organic farming. That is not so. While organic farming is all about maintaining soil health ('feeding the soil' rather than 'feeding the plant'), it entails producing the highest yields possible, in a sustainable, eco-friendly manner using a number of techniques. Indeed, for any agriculture system to succeed, it must take a responsibility towards feeding the soil. The soils which are depleted and infertile have to be made 'alive' with adequate humus or organic matter content. Depending on the current status by scientific assessment, chemicals may be required to balance the nutritional status in the initial years. But in the longer run, they may have to be completely avoided. Similarly, external inputs, even if eco-friendly or bio-degradable are ecologically questionable and commercially unsustainable for an organic farm need to be avoided. Work is hence required to produce on farm inputs such as composts and biopesticides.

Organic farming entails a lot of hard work and even systems like Natural farming - as propagated by Fukuoka of Japan - entail tremendous understanding of nature, natural agriculture, the local ecosystem.

Questions

1. What is Organic Farming?
2. Give the USDA definition of organic farming.
3. Mention the main concepts of organic farming.
4. Explain the "Healthy Plant Concept" of organic farming.
5. Explain the 'Living Soil Concept' of organic farming.
6. Mention the main principles of organic farming.
7. What are the characteristics of an organic farm?
8. What do you meant by pest resurgence?
9. Explain the development of pest resistance against chemicals.
10. What are the beneficial effects of humus in the soil?
11. What is the role of organic farming according to the principle of health?
12. Why people choose organic food?
13. Explain Principle of fairness.
14. State the Principle of care.
15. What are the basic differences between conventional and organic farming?
16. What are the specialities of traditional farming?
17. What are the benefits of natural farming?
18. What is Permaculture?

19. Explain the following terms in relation to organic farming

    a. Traditional Farming

    b. Bio-dynamic Farming

    c. Sustainable Agriculture

20. Fill in the blanks :

    a) Organic farming includes all agricultural systems which promotes production practices that are ...., socially and ....sound.

    b) Farming is the ....of man in the ....of plants, animals or birds to get utility products of our choice.

    c) Use of .... and .... has increased farm production which resulted in Green revolution.

    d) In organic farming, ... is considered as the key to successful production.

    e) The term Organic farming is derived from the concept of.......................................

    f) "Look to Land" is a book written by ....

    g) Organic Farming emphasizes the use of management practices in preference to the use of ....inputs.

    h) The productive capacity of soil is linked with its ....which in turn is influenced by ...., ... content of soil.

    i) .... is the most dynamic part of a living soil.

    j) ....content in the soil minimizes the leaching loss of nutrients.

    k) In modem farming, ...has displaced the diverse and locally adapted indigenous cultivars.

    l) Regular replenishment of ....and its maintenance at optimal level are of primary importance in sustaining soil productivity.

    m) Organic farming is a process based farming where the .... of produce is given more importance.

    n) According to the principles of organic farming, ... is the wholeness and integrity of living systems?

o) Organic farming is intended to produce ....and .... food.

p) According to the principle of ecology, production is to be based on ....and..........................

q) Principles of fairness states that organic farming aims to produce a ....supply of ....food.

r) ....and...............................are the key concerns in management, development and technology choices in organic agriculture.

s) According to the principles of organic farming, ...is given central importance.

t) Traditional farming considers Earth as a.............................

u) In traditional farming, ... were used as indicators of weather, water, minerals etc.

v) .... is a holistic system of agriculture that seeks to connect nature with cosmic creative forces.

w) .... is the man behind the development and growth of Biodynamic farming.

x) Natural Farming is also known as................ or ....

y) Permaculture basically means a system of agriculture which relies on ....crops.

z) LEISA is ....

21. Expand the following:

IFOAM, FAO, WHO, WTO, HYV, GM crops.

Chapter - 2

Scope of Organic Farming in India

The need for organic farming in India arises from the unsustainability of agriculture production and the damage caused to ecology through the conventional farming practices. The present system of agriculture which we call '*conventional+*' and practiced the world over evolved in the western nations as a product of their socio-economic environment which promoted an over riding quest for accumulation of wealth. This method of farming adopted by other countries is inherently self destructive and unsustainable.

The modern farming is highly perfected by the Americans who dispossessed the natives of their farms right from the early period of the new settlers in US. The large farms appropriated by the immigrants required machines to do the large scale cultural operations. These machines needed large amount of fossil fuels besides forcing the farmers to raise the same crops again and again, in order to utilize these machines to their optimum capacities. The result was the reduction of bio-diversity and labour. The high cost of the machines necessitated high profits, which in turn put pressure to raise productivity. Then, only those crops with high productivity were cultivated which needed increased quantities of fertilizers and pesticides. Increasing use of pesticides resulted in the damage to environment and increased resistance of insects to them. Pesticides harmed useful organisms in the soil.

The monoculture of high yielding seeds required external inputs of chemical fertilizers. The fertilizers also destroy soil organisms. They damage the rhizobia that fix nitrogen and other micro organisms that make phosphates available to plants. The long term effect was reduction of crop yields. The damaged soil was easily eroded by wind and water. The eroding soil needed use of continuously increasing quantities of fertilizers, much of which was washed/leached into surface and underground water sources.

The theme of consumer welfare has become central in the economic activities in the developed countries in the world. Sustainable agriculture based on technologies that combine increased production with improved environmental protection has been accepted as absolutely essential for the maximization of the consumer welfare. The consumers are increasingly concerned about the quality of the products they consume and food safety has become a crucial requirement. Safety, quality and hygienic standards are increasingly being made strict. The mad cow disease and the question of genetically modified food production are the recent instances, which made the countries to tighten the laws. Mycotoxin contamination, unacceptable levels of pesticide residues and environment degradation are the problems on which the attention is centred. Keeping the interests of the consumers, the European Union has taken tough measures including criminal prosecution to ensure food safety. Another area to increase the consumer welfare is promotion of the eco-friendly methods in agriculture. No-till, or conservation agriculture, lower input approaches of integrated pest or nutrient management and organic farming are some of them.

The Indian agriculture switched over to the conventional system of production on the advent of the green revolution in the 1970s. The change was in the national interest which suffered set backs because of the country's over dependence on the foreign food sources. The national determination was so intense that all the attention was focused on the increase in agriculture production.

The agriculture and allied sectors in India provide employment to 65 per cent of the workers and accounts for 30 per cent of the national income. The growth of population and the increase in income will lead to a rise in demand for food grains as also for the agricultural raw materials for industry in the future. The area under cultivation, obviously, cannot be increased and the present 140 million hectares will have to meet the future increases in such demands. There is a strong reason for even a decline in the cultivated area because of the urbanization and industrialization, which in turn will exert much pressure on the existing, cropped area.

Science and technology have helped man to increase agricultural production from the natural resources like land. But the realization that this has been achieved at the cost of the nature and environment, which support the human life itself, is becoming clear. It has been fully evident that the present pattern of economic development, which ignores the ecology and environment, cannot sustain the achievement of man without substantial erosion of the factors that support the life system of all living things on the Earth. The evidence of the ill effects of development is well documented. As said earlier, we in India have to be concerned much more than any other nation of the world as agriculture is the source of livelihood of more than 6-7 million of our people and it is the foundation of the economic development of the country.

There were times when people lived close to nature with access to flora and fauna in healthier and cleaner surroundings. One has to look back at our present metropolitan cities or other large towns before the past fifty years as recorded in history of the present elder generation to see the striking differences in the surroundings in which the people lived there. Land, water and air, the most fundamental resources supporting the human life, have degraded into such an extent that they now constitute a threat to the livelihood of millions of people in the country.

Ecological and environmental effects have been highly publicised all over the world. Many times, these analyses have taken the shape of doomsday forecasts. Powerful interests in the developed western countries have also politicised these issues to take advantage of the poor nations of the world. Efforts to impose trade restrictions on the plea of environment protection are a direct result of these campaigns. But we have to recognize that the abysmal level to which we have degraded our resources, requires immediate remedial measures without terming the demand for them as the ploys of the rich nations to exploit the poor.

Another turn of the events has been the blame game for ecological problems stated at the Earth Summit and other international conferences. The developed countries, it is true, are to a great extent instrumental to degrade the environment. However, the poorer countries of the world including India cannot delay or ignore the need for remedial measures, which are to be effectively implemented. We cannot gloss over the fact that we have also contributed to the degradation of ecology; look at the droughts and floods, disappearance of forests, high noise level and air pollution in the cities which are our own creations.

Organically cultivated soils are relatively better attuned to withstand water stress and nutrient loss. Their potential to counter soil degradation is high and several experiments in arid areas reveal that organic farming may help to combat desertification. It is reported that about 70 hectares of desert in Egypt could be converted into fertile soil supporting livestock through organic and biodynamic practices. India, which has some areas of semi-arid and arid nature, can benefit from the experiment.

The organic agriculture movement in India received inspiration and assistance from IFOAM which has about 600 organizational members from 120 countries. All India Federation of Organic Farming (AIFOF) is a member of IFOAM and consists of a number of NGOs, farmers' organisations, promotional bodies and institutions.

The national productivity of many of the cereal crops, millets, oilseeds, pulses and horticultural crops continues to be one of the lowest in the world in spite of the green revolution. The fertilizer and pesticide consumption has increased manifold; but this trend has not been reflected in the crop productivity to that extent. The country's farming sector has started showing indications of reversing the rising productivity as against the increasing trend of input use.

The unsustainability of Indian agriculture is caused by the modern farming methods which have badly affected/damaged production resources and the environment.

2.1 Affects of Modern Farming Technology

The role of agriculture in economic development in an agrarian country like India is a pre-dominant one. Agriculture provides food for more than 1 billion people and yields raw materials for agro-based industries. Agricultural exports earn foreign exchange. Modernization of Indian agriculture began during the mid-sixties which resulted in the green revolution making the country a food grain surplus nation from a deficit one depending on food imports. Modern agriculture is based on the use of high yielding varieties of seeds, chemical fertilizers, irrigation water, pesticides, etc., and also on the adoption of multiple cropping systems with the extension of area under cultivation. But it also put severe pressure on natural resources like, land and water. However, given the continuous growth of modern technology along with the intensive use of natural resources, many of them of non-renewable. It is felt that agriculture cannot be sustainable in future because of the adverse changes being caused to the environment and the ecosystem. The environmental non-degradable nature of the agricultural development and its ecological balance have been studied in relation to the modem Indian farming system by experts which shows exploitation of land and water for agriculture, and the excessive use of chemicals.

2.1.1 Chemical contamination

Fertilizers: India is the third largest producer and consumer of fertilizers in the world after China and the USA. Consumption of chemical fertilizers (N,P,K) has been increasing in India during the past thirty years at a rate of almost half a million tonnes on an average, a year. It was only 13.13 kg/ha in 1970-71, 31.83 kg/ha in 1980-81, 74.81 kg/ha in 1995-96 and 96 kg/ha in 1999-2000. It shot up to about 116.5 kg/ha with total nutrient consumption of 22.57 million tonnes during 2007-08. The consumption of major fertilizers namely, Urea, DAP, MOP, SSP and Complexes were 25.96, 7.50, 2.88, 2.29 and 6.57 million tonnes.

The present use of about 128.3 kg (2012-13) of fertilizers per ha in India appears to be modest compared to the advanced countries. Currently about 80 per cent of the fertilizer is consumed in only about 120 districts constituting less than 33 per cent of the gross cultivating area. Experts point out that the efficiency of fertilizer use in India is only 30-35 per cent as the balance 65-70 per cent reaches the under ground water. The intensity of their use in a few regions and a few crops are causes of serious concern to human health, soil, water, environment and thus to the sustainability of agriculture production in the country.

It is true that the increasing use of fertilizer at high rates has boosted agricultural production in the country. But it has also caused adverse impact on soil and water as well as environment. Several studies on the effects of high level of fertilizer application on soil health have confirmed the adverse impacts.

Both drinking and irrigation water wells in large numbers have been found contaminated with nitrates, some of them are having even 45 mg per litre, well above the safe level.

Long term continuous use of high doses of chemical fertilizers badly affects the physical, chemical and biological properties of the soil. A study at the University of Agricultural Sciences, Bangalore confirmed the deterioration of soil health because of the reduction in water holding capacity, soil pH, organic carbon content and the availability of trace elements such as zinc in case of ragi crop even with the application of normal doses of fertilizer in the long run (Hegde, et. al., 1995). In the long run, increasing nitrogenous fertilizer use leads to the accumulation of nitrates in the soil. The application of sulphatic fertilizers leaves sulphates in the soil. Rainfall and excessive use of irrigation water cause these chemicals to change the alkaline or acidic nature of the soil. The nitrates go to the rivers, wells, lakes etc. And also leak into the drainage system which goes into the drinking water contaminating the environment. It also causes depletion of the ozone layer adding to the global warming. Use of nitrogen in the form of ammonium sulphate in the rice crop emanates ammonia polluting the atmosphere. The heavy metals present in the fertilizers and sewage sludge leach into ground water. Table 2.1 shows the content of some heavy metals in fertilizers and sludges.

Table 2.1 : Content of Heavy Metals in Fertilizers and Sludges

(metals: mg/kg)

Source	Cd	Cr	Cu	Pb	Zn
Ammonium Nitrate	1.1	2.5	3.6	5.4	11.7
Super Phosphate	16.6	157.0	22.6	20.6	244.0
NPK Fertilizers (8-10-18)	4.9	54.3	8.3	3.2	97.5
Sewage Sludge	20.0	500.0	250.0	700.0	3000.0

Source : Deb and Joshi (1994)

Pesticides: The use of chemical pesticides began with the discovery of toxicological properties of DDT and HCH during the Second World War. Many chlorinated hydrocarbon insecticides like aldrin, dieldrin, toxaphane, chlordane, endosulfan, etc. came into the market during the second half of the last century.

Simultaneously, organophosphate and carbonate compounds were employed in agriculture. A new group of insecticides, such as premethrin, cypermethrin, fenalerate, etc which were effective at low doses came into being in the 1970s.

The use of pesticides has helped in increasing agriculture production and also led to the development of resistance in pests, contamination of the environment and resurgence of many pests.

There are about 1000 agrochemicals in use in the world over. India accounts for about 3.7 per cent of the total world consumption. At present, our consumption is about 90,000 tonnes of plant protection chemicals. It comes to about 580 grams per ha compared to 10-12 kg/ha in Japan and 5 kg/ha in Europe. However, the use of pesticides in India is uneven like the fertilizers. While in cotton it is about 3 to 4 kg/ha, in pulses it comes to below 500 grams/ha (Kathpal and Beena Kumari, 1997). Pesticide application is also concentrated in some areas as in the case of fertilizers mentioned earlier.

Agricultural chemicals have become a major input in Indian agriculture with the increasing demand for food, feed and fibre. The pesticide consumption was about 2000 tonnes annually during the 1950s. India happens to be the second largest manufacturer of pesticides in Asia after Japan. It is also of interest to know that in spite of increased consumption of plant protection chemicals, the produce loss due to insects and pests increased by 5 times during the period from 1988 to 1995.

Increasing application of fertilizer also leads to increasing use of pesticides to control pests and diseases. The trend of increasing fertilizer use also compels the farmers to enhance the use of pesticides as well. For example, the use of fertilizers in increasing amounts leads to growth of weeds and in the process of weedicide use many plants growing nearby also get killed, which reduces the biodiversity. Meanwhile, the weeds also develop resistance to herbicides and the quest to formulate even powerful herbicides begins.

Consumption of pesticides increased from 24.32 thousand tonnes in 1970-71 to 75 thousand tonnes in 1990-91 and it slowed down during the subsequent period. In the year 2005-06, consumption of pesticides in agriculture was 44177 tonnes. Insects, pests and diseases like viral, bacterial and fungal affect the high yielding varieties of crops. Almost all pesticides are toxic in nature and pollute the environment leading to grave damage to ecology and human life itself. This indiscriminate use leaves toxic residues in food grains, fodder, vegetables, meat, milk, milk products, etc besides in soil and water.

High doses of pesticides severely affect the aquatic animals, fish and the wild life. Insects develop resistance to insecticides in crops like cotton and in turn force the farmers to the excessive use of them. Cases of pesticide poisoning and human and animal deaths are also reported. Pesticides irritate the skin and the respiratory system in the humans gets damaged.

It was found that all water bodies like, rivers, canals, lakes, tanks and ponds and also the costal water were contaminated with high amounts of DDT, HCH and other organochlorine pesticides. River water is seen as more contaminated than other water sources.

Contamination of drinking water with DDT and HCH is reported from different States. Since the concentrations of contaminants are higher than MRL (0.5 ppb) values fixed by the Environmental Protection Agency (EPA), the seriousness of the problem can be gauged. Pesticides also contaminate animal feeds and fodder. Green fodder, paddy and wheat straw contain residues of DDT and HCH.

Milk and milk products are also affected by the pesticide use. Both bovine and human milk showed high levels of pesticide contamination. The sources of contamination of bovine milk are traced to the fodder and feed concentrates and in case of the human milk, the consumption of contaminated food by the lactating mothers is reported to be the reason.

Infant formula/baby milk powders also showed DDT and HCH contamination level ranging from 94 to 100 per cent. Butter and ghee, the other animal products revealed high contamination levels in many parts of the country.

Cereals like wheat and rice were seen contaminated highly by pesticides like, DDT, HCH and Malathion. The case of vegetable, vegetable oil, honey, fish, etc is also not different as they too have unacceptable high pesticide residue content levels.

The adverse effect of pesticide contamination on humans in India is understood from the study of dietary intake. Such studies, although a few in number, have confirmed high levels of pesticides (mainly, DDT and HCH) contamination which come to more than 3 to 5 times than the agriculturally developed countries.

The daily intake of pesticide per individual is estimated to be about 0.51 milli-grams which is above the accepted level. The Indian Institute of Horticulture Research has reported contamination of 50 per cent of the fruits and vegetables sold in the Bangalore market with the residues of DDT and HCH (Prakash, 2003).

Use of herbicides over period results in the shift of the weed flora. The weeds of minor importance, often, become major weeds. Repeated application of weedicide helps the development of resistance in weed at alarming proportions. The remedy recommended is rotation of herbicides or the use of other herbicides. The end result is contamination of ground water and soils inflicting damage on environment.

The number of herbicides registered in India comes to about 28 in 1997-98 which was only 10 before 10 years. This is often compared to about 300 herbicides available in the North America. There are only 11 herbicides manufactured in the country and the herbicides consumption was about 6310 tonnes during 2006-07. It is reported that the level of herbicide use in rice, wheat, and tea in India is almost the same that of the world at large. Sugarcane, soyabean, groundnut, coffee, cotton, onion and potato are the other crops, which find widespread application of herbicides in India.

The contamination of water, air and soil with toxic synthetic fertilizers, pesticides and herbicides leads to increasing deaths of many creatures, and to human illness and mortality. The end result is loss of biodiversity and natural harmony, increased expenditure to purify water, air, etc. The toxins in the food crops cannot be removed and the threat to human existence itself seems to be real.

2.1.2 Salinity and waterlogging

Water is one of the important inputs for the vigorous growth and high yields of crops. The modernisation of Indian agriculture has resulted in the increased use of irrigation water. The area under irrigation has grown substantially during the past four decades.

In 1951, only 22.6 million hectares of land was under irrigation. By 1985-86, land under irrigation nearly increased to 58.82 million hectares. By 2000-01, another 21.24 million hectare land were brought under irrigation, carrying total 80.06 million hectares. This may be judges against the whole potentiality of 113 million hectares by 2010 A.D. At present around 28 per cent of the net sown area is under irrigational use i.e. 63 million hectares, although the gross irrigated area is approximately 89 million hectares. However, not more than 50 percent of the net sown area will eventually be brought under irrigation.

Heavy irrigation is necessary to get high production, as the new varieties cannot withstand water scarcity. This leads to salinity and water logging leaving the land uncultivable. Over exploitation of underground water is another effect. When water table falls, increasing energy will be required to lift water for irrigation.

Irrigation is necessary for the vigorous growth and high yields of crops in the modern method of cultivation. Many of the crops, particularly the rice and wheat high yielding varieties need more irrigation water than the traditional varieties. The area under irrigation in the country is only about 35 per cent and the remaining is still dependent on rains. So, there is a necessity to use irrigation water judiciously. Its excessive use results in severe ecological dangers like water logging of vast cultivated areas by seepages from canals. The loss of

water through seepages and evaporation is estimated to be about 38 per cent. Flooding also results in run off and leaching losses of fertilizer nutrients, pesticides and soil particles. Excessive use of canal water makes the field vulnerable to soil erosion. The excessive irrigation in certain areas results in wastage as evident from the water logging of vast cultivated areas caused by the seepage from many water sources.

Water logging is harmful to the soil. Seepage of canal water leads to salts present in the lowest layer of soil come up to the surface and the soil may turn alkaline or saline. Dams and multipurpose projects degrade the soil in the command area due to soil salinity and water logging. The Chambel region in Rajasthan and Madhya Pradesh, the command area of the Bhakra Nangal, etc are the examples of water logging created by huge water irrigation projects. Crops irrigated by sewage water have adverse effects on the health of the human population consuming the produce. The workers work on these farms also face health hazards.

2.1.3 Depletion of energy resources

Chemical fertilizers, pesticides, herbicides, etc are manufactured using the non-renewable materials like the fossil fuels. The global demand for oil and natural gas is increasing and thus the price of the inputs to agriculture is bound to rise. India's petroleum resources, which presently meet only about 30-35 per cent of the consumption demand, are under pressure. Increasing demand for chemicals and energy in agriculture sector will have affects on our energy sources.

The investments in agriculture have to be increased to meet the rising input costs and larger areas are brought under farming to earn profits. Large farms have to transport the produce to distant areas. Again, energy will be required for transportation, processing and packaging.

The rice-wheat cropping pattern and the cultivation of crops like sugarcane require high irrigation, which results in the depletion of water level. It was found that the water level in the states of Punjab and Haryana had gone down by 0.3 to 1 m per annum during a period of 10 years due to the excessive use of water for paddy crop.

2.1.4 Input-Output imbalance

A crop, in its growth process, incorporates a part of the soil fertility into the parts of the plant. The roots remain in the soil. The leaves and stems are fed to the cattle/burnt as fuel/directly returned to the soil. The consumed part by cattle and human also go back to the soil. The practice of commercial farming leads to continuous export of the soil fertility to outside the farming areas as the organic matter leaves the locality. The soil nutrients in the form of farm produces continue to be exported. The import of chemical fertilizers cannot compensate the loss of soil nutrients through exports. The soil becomes powdery and gets eroded by wind or rain. If the harvests are exported from the country, the loss will be higher.

The Earth can produce only a limited amount of biomass from a given area. If man tries to extract more, the system degenerates. If we realize that we live in a closed eco-system and act in tune, the resources can be used in a sustainable way. The maintenance of a dynamic equilibrium balances the inputs and outputs in a region. If agriculture becomes unsustainable, no society can survive for a long time.

2.1.5 Reduction in genetic diversity

The genetic base of crops is very important and a reduction of genetic diversity leads to the emergence of pests on a large scale. Farmers, in olden times, apart from using the crop rotation methods to maintain the soil fertility also relied on the genetic means to increase crop production. Relying exclusively on nation's own reserves of fertility and immunology, the farming community by evolving trial and error methods discovered hybrid varieties of crops by crossing the related strains. These crosses were from the same environment and no violence was used to separate them from nature by maintaining the ecological balance (Alvares, 1996).

The high yielding varieties of crops are the crosses from different environments and distantly related strains. For example, the high yielding rice variety got by the crossing between the dwarf and non-dwarf varieties has major genetic weaknesses. The dwarf gene is susceptible to pest and viral attacks and the seed cannot manifest its potential without chemical fertilizer. Thus, an artificial environment has to be created for the growth of the crop.

Thus synthetic fertilizers suppress natural fertility, which results in larger population of pests. The new technology adopted then depends upon the replacement of the local/traditional varieties of seeds. But this results in the reduction of genetic diversity and increase in genetic erosion.

2.1.6 Low productivity

The productivity of cereals, millets, oilseeds, pulses and plantation crops is very low in comparison with those in other countries in the world. This is in spite of our success in improving the quality of seeds and adoption of efficient technology. The impact of green revolution is showing signs of weakness and production appears to have decreased even after an increase in the inputs used (Veeresh, 1999).

The production of food grains in the country increased very substantially during 1960 to 1980 to reach 160 million tonnes from 60 million tonnes. But the decade ending 1990 and 2000 did not witness such increases and the attainment of the targeted production of 280 million tonnes to meet the demand of the population by 2020 seems to be difficult.

The reasons attributed to the low productivity are the drastic reduction in soil nutrients in the areas where fertilizer is used intensively in which the organic matter is not supplemented.

The relevance and need for an eco-friendly alternative farming system arose from the ill effects of the chemical farming practices adopted worldwide during the second half of the last century. The methods of farming evolved and adopted by our forefathers for centuries were less injurious to the environment. People began to think of various alternative farming systems based on the protection of environment which in turn would increase the welfare of the humankind by various ways like clean and healthy foods, an ecology which is conducive to the survival of all the living and non-living things, low use of the non-renewable energy sources, etc. Many systems of farming came out of the efforts of many experts and laymen. However, organic farming is considered to be the best among all of them because of its scientific approach and wider acceptance all over the world.

2.2 Benefits of Organic Farming

Organic agricultural practices are based on a maximum harmonious relationship with nature aiming at the non-destruction of the environment. The developed nations of the world are concerned about the spreading contamination of poisonous chemicals in food, feed, fodder and fibre. Naturally, organic farming system is looked upon as one of the means to remedy these maladies there. Organic farming aims at reducing the costs of production and helps the farmers to get reasonable returns. Organic farming is today's answer not only to sustained productivity, but also to safe and nutritious food, which is increasingly demanded by conscious consumers within and outside India. Organic products also promise better prospects for market and trade. There is an ever growing export market for organic products. Similarly the domestic market is also emerging. Products that are certified and sold as 'organic' can fetch a premium price compared to conventional products.

Healthy foods: A study conducted in USA on the nutritional values of both organic and conventional foods found that consumption of the former is healthier. The organic food, in general, had more than 20 per cent less of the bad elements and about 100 per cent more of the good elements.

Improvement in soil quality: Soil quality is the foundation on which organic farming is based. Natural plant nutrients from green manures, farmyard manures, composts and plant residues build organic content in the soil. It is reported that soil under organic farming conditions had lower bulk density, higher water holding capacity, higher microbial biomass carbon and nitrogen and higher soil respiration activities compared to the conventional farms (Sharma, 2003). This indicates that sufficiently higher amounts of nutrients are made available to the crops due to enhanced microbial activity under organic farming.

Increased crop productivity and income: Results reported from 1050 field demonstration cum trials under the National Project on '*Development and Use of Biofertilizers*' in different parts of the country show an increase of 4 per cent in yield in plantation crops, 7 per cent in fruit crops, 9 per cent in wheat and sugarcane, 10 per cent in millet and vegetable, 11 per cent in fibre, condiments and spice crops, 14 per cent in oilseeds and flowers (Bisoyi, et. al., 2003).

A study of 100 farmers in Himachal Pradesh during a period of 3 years found that the total cost of production of maize and wheat was lower under organic farming and the net income was 2 to 3 times higher. Both productivity and premium prices contributed to the increased profitability. Another study of 100 farmers of organic and conventional methods in five districts of Karnataka indicated that the cost of organic farming was lower by 80 per cent than that of the conventional one (Thakur, et. al., 2003).

Low incidence of pests: The study of the effectiveness of organic cotton cultivation on pests at the farm of Central Institute for Cotton Research, Nagpur revealed that the mean monthly counts of eggs, larva and adults of American Bollworm were far lesser under organic farming than under the conventional method (Sharma 2003).

Bio-control methods like the neem based pesticides to Trichoderma are available in the country. Indigenous technological products such as Panchagavya (five products of cow origin) which was experimented at the University of Agricultural Sciences, Bangalore found to control effectively wilt disease in tomato (Prakash 2003).

Economic Benefits: Reduction in the use of external inputs and increase in output of organic produces with greater potential to benefit the health of farmers and consumers. More Productivity through the incorporation of natural process like natural cycles, nitrogen fixation and pest-predator relationship into the agricultural production process. Greater productive use of the biological and genetic potential of plant and animal species. Long term sustainability of production levels. Profitable and efficient production with emphasis on improved management and conservation of soil, water, energy and biological resources.

Ecological Benefits: Organic farming is much better for the environment as the energy consumption is much less than in the chemical farming. It also uses less manurial inputs and completely avoids the synthetic fertilizers which otherwise pollute the soil, water and air. It promotes biodiversity and a great variety of animals and plant interaction on earth. Organic farmers focus on preserving the habitats of all species and their surrounding environment (including the air and water). Organic farming releases much less carbon dioxide than other farming system.

Social Benefits: Organic farming practices can be adopted in small farms and benefit marginal farmers. It could reduce dependency on external inputs and costly technologies thus reducing the competitiveness and disparity among the farmers in a community. It will also lead to food security at the family level and national level. Organic farming is revival of a culture and brings back the indigenous knowledge, beliefs and value system that are almost on extinction now. It also contributes to employment generation at the community level.

Employment opportunities: According to many studies, organic farming requires more labour input than the conventional farming system. Thus, India which has a very large amount of labour unemployment and under employment will find organic farming an attraction. Moreover, the problem of periodical unemployment will also get mitigated because of the diversification of the crops with their different planting and harvesting schedules resulting in the requirement of a relatively high labour input.

Indirect benefits: Several indirect benefits from organic farming are available to both the farmers and consumers. While the consumers get healthy foods with better palatability and taste and nutritive values, the farmers are indirectly benefited from healthy soils and farm production environment. Eco-tourism is increasingly becoming popular and organic farms have turned into such favourite spots in countries like Italy. Protection of the ecosystem, flora, fauna and increased biodiversity and the resulting benefits to all human and living things are great advantages of organic farming which are yet to be properly accounted for.

2.3 The International Scene of Organic Farming

The negative effects of modern chemical based farming system were first experienced by those countries, which introduced it initially. So, naturally, it was in those countries organic farming was adopted in relatively large scales. There are very large organisations promoting the organic farming movement in European countries, America and Australia etc. These organisations, for example, the International Federation of Organic Agriculture Movements (IFOAM) and Greenpeace have studied the problems of the chemical farming methods and compared the benefits accruing to the organic faming with the former. Organic farming movements have since spread to Asia and Africa too.

IFOAM was founded in France in 1972. It spearheads and coordinates organic farming efforts the world over by promoting organic agriculture as an environment friendly and sustaining method. It focuses on organic farming by highlighting the minimum pollution and low use of non-renewable natural resources through this method. It has about 600 organisational members spread over about 120 countries including India.

IFOAM undertakes a wide range of activities related to organic farming such as exchanging knowledge and thoughts among its members; representation of the movement in governmental, administrative and policy making forums in the national and international arena; updating of production, processing and trading standards; formulation and coordination of research projects; and holding of international conferences and seminars. IFOAM participates in the activities related to organic farming under the auspices of the United Nations and keeps active contacts with several international NGOs.

The Food and Agriculture Organisation (FAO) of the United Nations provides support to organic farming in the member countries. It also attempts the harmonization of national organic standards, which is absolutely essential to increase international trade in organic products. The FAO has, in association with the World Health Organisation (WHO), evolved the Codex Alimentarius for organic products.

Organic farming has several advantages over the conventional one apart from the protection of both the environment and human health. Improved soil fertility, better water quality, prevention of soil erosion, generation of rural employment, etc. are some of them.

The concept of quality food has undergone a drastic change over the past few decades. It does give emphasis on the characteristics of the end product, but the process and method of production and transport are now considered equally important. Not only the importers but also the domestic retailers have their own quality specifications or standards, which in many cases are tougher than those of the government regulations. Consumers have become health conscious and are willing to pay for the clean, healthy and natural food. Many developed countries have various support programmes to help organic farming with financial incentives and technical guidance.

Organic farming is gaining gradual momentum across the world. Based on the global survey on organic farming carried out in 2007-08 by the Research Institute of Organic Agriculture (FiBL), the International Federation of Organic Agriculture Movements (IFOAM) and Foundation of Ecology and Agriculture (SOEL), organic agriculture is now practised in more than 130 countries with a total area of 30.4 million hectares in 0.7 million number of organic farms. This constitutes about 0.65% of total agricultural land of the world (Table 2.1).

Table 2.1 : Land area of major countries under organic farming (2007–08)

Country	Area under organic agriculture	Percentage of total agricultural land	Number of organic farms
Australia	12,294,290	2.8	1550
China	2,300,000	0.4	1600
Argentina	2,220,489	1.7	1486
USA (2005)	1,620,351	0.5	8493
Italy	1,148,162	9.0	45,115
Uruguay	930,965	6.1	630
Spain	926,390	3.7	17,214
Brazil	880,000	0.3	15,000
Germany	825,539	4.8	17,557
UK	604,571	3.8	4485
Canada	604,404	0.9	3571
France	552,824	2.0	11,640
India	528,171	0.3	44,926
World Total	**30,418,261**	**0.65**	**718,744**

Source : Willer, H. (2008)

Global demand for organic products remains robust, with sales increasing by over US$ 5 billion a year. Organic Monitor estimates international sales to have reached US$ 38.6 billion in 2006, double that of 2000, when sales was at US$ 18 billion. Consumer demand for organic products is concentrated in North America and Europe; these two regions comprise 97% of global revenues. Asia, Latin America and Australia are important producers and exporters of organic foods. A worldwide organic agriculture is presented in Table 2.1.

In India (during 2007-08), about 528,171 hectare area is under organic farming (this includes certified and area under organic conversion) with 44,926 number of certified organic farms. This accounts for about 0.3% of total agricultural land. Indian organic farming industry is estimated at US$ 78 million and is almost entirely export oriented. According to Agricultural and Processed Food Products Export Development Authority (APEDA), a nodal agency involved in promoting Indian organic agriculture, about 585,970 tonnes of organic products worth of Rs 301 million are being exported from India. Growing awareness, increasing market demand, increasing inclination of farmers to go organic and growing institutional support have resulted in more than 200% growth in certified area during the last two years. The Indian organic foods industry is expected to witness a boom period. There is huge export opportunity for the Indian organic farming industry both in niche markets as well as mainstream market world over. The present status of organic production in india is given below :

National Programme for Organic Production (NPOP)

Present Status in India (2013-14)	
Total production	: 1.24 million MT
Total quantity exported	: 194088 MT
Total quantity exported realization	: 403 million US $
Total area under organic cultivation	: 4.72 million ha
Number of items exported	: 86

The major organic products sold in global market include Basmati rice, pulses, honey, tea, spices, coffee, oil seeds, fruits, processed food, cereals, herbal medicines and there value added products. The production is not limited to the edible sector but also produces organic cotton fibre, garments, cosmetics, functional food products, body care products, etc. Organic products are mainly exported to EU, US, Australia, Canada, Japan, Switzerland, South Africa and Middle East.

2.4 Relevance of Organic Farming in Indian Context

In many countries, organic agriculture got promoted as a reaction to the high external input based techniques ushered in during the green revolution. Consumer demand for better quality food also helped it. In the Indian context, the following benefits are considered important for the future of organic agriculture development.

- The irrigation, new seeds and fertilizer required high initial investments and, thus, are beyond the reach of majority of the small farmers in India. The emphasis on organic practices involving use of techniques such as nitrogen fixing crops and green manuring, recycling nutrients through composting, deep rooting plants, avoiding soil loss and locally developed pest control measures allows poor, risk-averse farmers to produce food and generate income for the families on sustainable basis.
- Pressure on the cultivated land has led to soil degradation. In many cases marginal lands have been put to cultivation by adding considerable quantities of external inputs. In large areas of cultivated land, environmental damage is being associated with 'modern' agricultural practices.
- The concern for ensuring and enhancing agricultural output with due respect to ecological limits is significant in organic agriculture. The

productivity of natural resources and income distribution in rural areas are interrelated and impact on poverty. Wherever lands have become unproductive, it has led to income disparity and social concern. Farmers who opted for organic production system would be free from buying expensive inputs.

- Some super markets in India have started stocking organic products. There are organic and safe food bazaars coming up in metros and other cities. This fast pace of rising demand of organic products will be instrumental in giving boost to organic agriculture development in India.
- The Small and marginal farmers engaged in rainfed farming produce a variety of crops that does not require much of chemical inputs. These farmers qualify themselves as organic producers. However they need to get educated on the scientific aspects of organic production to get their fields classified as pure organic farm.
- In many predominantly agricultural areas in India, land productivity is declining even with higher doses of external inputs. Such farmers on their own began exploring alternatives. Now organic farming is proved to be a viable option for them because of the increasing demands for organic food and official support to organic agriculture.
- Increasing number of middle class consumers and their willingness to pay a premium for quality food has generated new demand for value added products. Organically raised and processed food should find ready acceptance in the markets. This has created new opportunity for the agri-entrepreneurs and organic food processors.
- Globalization and World Trade Organization regulations are expected to increase competition. On the other hand it may also offer an opportunity to organic producers in India to tap the market. Indian farmers have distinct competitive advantages.

It is very difficult to provide scientific data on the yield or economic returns. However experiences of different organic farmers reveal that there are economic benefits, ecological benefits as well as social benefits in converting to organic farm.

2.5 Organic Farming Scenario in India

Organic Agriculture is not a new concept to India. At the beginning of the 19th century, Sir Albert Howard (1873-1947), one of the most important pioneers of organic farming, worked in India for many years, studying soil-plant interactions and developing composting methods. In doing so, he capitalized substantially on

India's highly sophisticated traditional agricultural systems, which had long applied many of the principles of organic farming (e.g. crop rotations with legumes, mixed cropping, botanical pesticides etc.). He was the principal figure in the early organic movement and acknowledged as the ***Father of Modern Organic Agriculture.*** He worked in India as Agricultural Advisor. He observed and supported the traditional Indian farming practices over that of modern agricultural science. Though he came to India to teach Western agricultural techniques, he found that the Indians could in fact teach him more. One important aspect he took notice of was the connection between **healthy soil** and the **healthy populations, livestock** and **crop** in villages.

Introduction of Green Revolution agricultural technology in the 1960s reached the main production areas of the country, there were still certain areas (especially mountain areas) and communities (especially certain tribes) that did not adopt the use of agro-chemicals. Therefore, some areas can be classified as *'organic zone'*. An increasing number of farmers have consciously abandoned agro-chemicals and now started produce crop organically, as a viable alternative to Green Revolution agriculture.

India is classified into 21 agro-ecological zones based on temperature, soil conditions, and rainfall. Hence, each zone has a comparative advantage for the production of different products, e.g. tea, spices and fruits in the north eastern region; spices and coffee in the southern region; rice, wheat and fruits in the northern region; cotton and herbs in central and western region. Products with potential in the domestic market are fruits, vegetables, rice and wheat. Products with potential in the export market are tea, rice, fruits, vegetables, cotton, herbs and spices. Further, India has time tested indigenous farming systems and use of indigenous technical knowledge in agriculture and allied sector.

India has also the following advantages:

- India is strong in high quality production of tea, spices, rice specialties, ayurvedic herbs etc.
- India has a rich heritage of agricultural traditions which are suitable for designing organic production systems.
- In several regions of India agriculture is not very intensive as regards the use of agro-chemicals (mountain areas).
- The labour is relatively cheap compared to agro-chemicals.
- The NGO sector in India is very strong and has established close linkages to large numbers of marginal farmers.

Certified organic farming in India is still in a nascent stage. There are numerous initiatives of farmer groups, NGOs and corporate bodies producing organically and selling the products on domestic and international markets. In addition to improved food quality and environment, the potential for organic agriculture to reduce production costs, to stabilize yields and to increase farmers' income, especially in marginal production regions, is considered as high under the Indian scenario. In 2000, the Indian Government started to promote organic agriculture on various levels under the National Programme for organic Production (NPOP).

According to official statistics, until February 2001 there were only 304 organic farms in India and the figure has increased to 44,926 farms during 2007-08 (Table 2.1), accounting for barely 0.3% (528,171 hectares) of total agricultural land. However, the database is still very poor and it can be assumed that the real figures are much higher. Presently, organic production for the whole of India is around 1.24 million MT covering an area of 4.72 million ha (2013-14). Organic products produced in India (Table 2.2) are tea, spices, rice, coffee, cashew nuts, oil seeds, wheat, pulses, cotton, herbal extracts, fruits and vegetables. Considering the increase in organic farming activity in India, the major organic products for which a growing demand is anticipated would be tea, spices and fruits.

Table 2.2 : Major products produced in India by organic farming

Type	Products
Cereals	Rice, wheat
Spices	Cardamom, black pepper, white pepper, ginger, turmeric, vanilla, mustard, tamarind, clove, cinnamon, nutmeg, chilly
Pulses	Red gram, black gram
Fruits	Mango, banana, pineapple, passion fruit, sugarcane, orange, cashew nut, walnut
Vegetables	Okra, brinjal, garlic, onion, tomato, potato
Oil seeds	Sesame, castor, sunflower
Beverages	Tea, coffee
Others	Cotton, herbal extracts

2.5.1 Indian organic products in the markets

The domestic market for organic products is not yet as developed as the export market. The products available in the domestic market in organic quality are rice, wheat, tea, coffee, pulses, fruits and vegetables. As most organic production originates from small farmers, wholesalers and traders account for a 60% share in the distribution of organic products. Large organized producers distribute their products through supermarkets as well as through self-owned stalls.

Considering the profile of existing consumers of organic products, supermarkets and restaurants are the major marketing channels for organic products. Khadi and Village Industries Commission (KVIC) with its 7000 outlets in India has taken the initiative to market organic products in the domestic market. In 2002, KVIC successfully launched its organic product line 'Desi Ahaar' in some of its outlets – and was sold out in short time. Besides KVIC there are a number of formal and informal initiatives to sell organic products in the domestic market. In the long term, the domestic market is expected to be much more important for Indian organic farmers than export markets.

Trade with organic products are showing growth rates of 15–20%, which are rarely found in food markets. Products with potential in the domestic market are fruits, vegetables, rice and wheat. Products with potential in the export market are tea, rice, fruits and vegetables, cotton, wheat and spices. Products for which production in India has a comparative advantage are given below:

- Basmati rice (if there is enough volume)
- Spices (from small quantities to 50 tonnes)
- Tea (only if high quality, up to 40 tonnes)
- Coffee (only if high quality)
- Mango (fresh, dry and puree, quantities up to 50 tonnes)
- Pineapple (juice, dry, small quantities up to 50 tonnes)
- Bananas (fresh and dry, small quantities up to 70 tonnes)
- Organic vanilla (only if good price, high quality)
- Cashew nuts (only if good price and high quality)
- Protein grains (if there is not a problem with the quality).

Organic agriculture is in many ways would be an eminently preferable model for the development of agriculture in India, especially where yields under conventional farming are low. Organic agriculture offers multiple benefits. These include economical aspects (e.g. price premiums, high demand), natural resource conservation (e.g. improved soil fertility and water quality, prevention of soil erosion, preservation of natural and agro-biodiversity) and social benefits (e.g. generation of rural employment and corresponding lower urban migration, improved household nutrition and local food security, reduced dependence on external inputs). Organic systems can achieve similar yield levels as conventional ones, and can even contribute to increase productivity in the long term, as experienced by many organic farmers.

Questions

1. What are the affects of Modern Farming Technology on Indian agriculture?
2. What are the benefits of organic farming?
3. How the use of chemical pesticides affects the economy security of farmers?
4. What are the major organic products sold in global market?
5. Explain the relevance of organic farming in Indian context?
6. What are the advantages for Indian farmers over their overseas counterparts?
7. Provide list of Indian organic products which has potential in the export market?
8. Explain the economic benefits of organic farming?
9. What are the ecological benefits of organic farming?
10. Mention the social benefits of organic farming?
11. Fill in the blanks :
    a. Father of Modern Organic Agriculture is.......................
    b. Long term continuous use of high doses of chemical fertilizers badly affects the ...., ... and .... properties of the soil.
    c. The contamination of ...., ... and ....with toxic synthetic fertilizers, pesticides and herbicides leads to increasing death of many creatures, and to human illness and mortality.
    d. Heavy irrigation is necessary to get high production, but this leads to .... and .... leaving the land uncultivable.
    e. ....a nodal agency involved in promoting Indian organic agriculture.
    f. India is classified into 21 agro-ecological zones based on ...., soil conditions, and ....
12. Expand the following:

    FiBL, APEDA, KVIC, NPOP, AIFOF

CHAPTER - 3

Organic Production System

Organic food production is as much a state of mind as it is a production system. The organic farmer must understand how everything is interrelated and how one set of circumstances will influence other factors in how the plant grows. A strong, healthy plant is much more important in an organic system then a conventional system. A two to three year leaning curve must be accepted by anyone intending to switch from conventional to organic farming. What works one year may fail the next.

Organic Crop Production

Biodiversity/ Sustainability	Natural Plant Nutrition	Natural Pest Management	Integrity
Rotation	Rotation	Rotation	Buffer
Green Manure	Green Manure	Green Manure	Records
Cover Crops	Animal Manure	Cover Crops	
Animal Manure	Composting	Composting	
Composting	Natural Fertilisers	Intercropping	
Intercropping	Foliar Fertilisers	Biocontrol	
Buffers	Tillage	Fire	
		Natural Pesticides	

An organic production system is designed to:

- enhance biological diversity within the whole system
- increase soil biological activity
- maintain long-term soil fertility
- recycle wastes of plant and animal origin in order to return nutrients to the land:

- rely on renewable resources in locally organized agricultural systems, thus minimizing the use of non-renewable resources
- promote the healthy use of soil, water and air as well as minimize all forms of pollution
- handle agricultural products with emphasis on careful processing methods in order to maintain the organic integrity and vital qualities of the product at all stages.

Organic farming is neither merely replacing the chemicals with organics nor it is going back to the traditional agriculture. It is the conversion of soil from non-living to living and to sustain the same. Normally it takes 3-4 years before the soil is converted to living. Once the soil is converted to organic, anything grown on such land the produce becomes organic provided the principles of organic standards are followed in the process of production, processing, transport and storage.

3.1 Main Requirements

There are specific requirements for most organically certified crops as well as livestock, fish farming, bee keeping, forestry and the harvesting of wild products. The main requirement in organic production are-

- Organic Standards requirements must be applied from the beginning of the organic production.
- Total crop production and animal husbandry should be converted to organic management as per Organic Standards which require a conversion period (or time that a farm has to use organic production methods before it can be certified, usually 2 - 3 years).

Other requirements of organic production are as follows:

Crop Production Requirements
• Selection of seeds and plant materials
• Maintenance of soil fertility and the recycling of organic materials
• Prohibition of genetically modified inputs
• Diversity of crops on farm
• Processing, packaging and traceability of products
• Use of organic fertilizers and compounds for the control of pests, diseases and weeds

Animal Production Requirements
• Animal health and welfare
• Nutrition and breeding
• Transport and slaughter procedures

The details of organic production requirement are discussed here.

3.1.1 Organic conversion period

- Duration of the conversion period is 24 months before the start of production cycle for annual crops, 36 months before harvest for perennial crops and 12 months for animal products. With regard to dairy & egg production and meat production this period is 30 days and 12 months respectively. All the organic standard ought to be met before sowing/planting of annual/perennial crop.
- Start of conversion period is calculated from the date of application to the certification body, when farmers commit themselves in following the standards.
- During conversion period, products can be labelled as "Produce of organic agriculture in the process of conversion" provided organic standards have been met for all at least 12 months.
- Converted land and animals should not get switched back and forth between organic and conventional management.

3.1.2 Organic crop production

Field history: At least 24-36 months of histories in terms of crops planted, crop yield, inputs applied are required for all fields. If farm is newly purchased or rented from previous farmers than get all supporting information about soil, plant or water tests, etc.

Following data on physico-chemical-biological environment should be collected:

Physico-chemical-biological environment

Climate	Agro-ecological zone, rainfall, temperature, wind, sunny days
Soil	Soil texture, soil pH, bulk density, colour, etc. at two depths: 10 and 15-30 cm; N, P, K, organic matter and heavy metals, C:N ratio, etc.
Topography	Slope and position on slope, (e.g. steep on top, medium in the middle, etc.), floodplain
Irrigation	Water source and quality (pH, heavy metals, D.O., TDS, EC, Arsenic, nitrates, fluoride, B.O.D., C.O.D., etc.), means and frequency of delivery, on-farm practices, water table, etc.
Farm site	Field history, crop varieties used, plant arrangement, fertiliser use, pest control, weeds, insects, diseases, pests, crop yields

Soil information: Soil and soil management is the foundation of organic production. Availability of all plant nutrients in an optimum concentration is essential for the normal growth and development of plants. Soil is the store-house for the supply of plant nutrients. Potential productivity of soils varies considerably because of differences in amount of plant nutrients. A detailed examination of plant nutrients (chemical properties) and physical properties and their correct evaluation would enable one to identify the constraints and to suggest the ways and means by which the inherently low productive soil could be made more productive. In actual practice, the physical and chemical analyses of soils indicate the conditions such as mechanical impedance, low water and nutrient retention, excessive and low permeability, acidity, alkalinity and lack of nutrients limiting crop production.

Following soil information should be collected for organic crop production:

- Examine the entire soil profile to ascertain probable root distributions, clay pans, high water table, or generally poor physical condition which may greatly influence the crop productivity under organic farming. Composite soil sampling and its laboratory analysis (physical, chemical and biological properties) of the organic production area should be done to determine the proper crop management practices for organic conversion.
- Each crop grown depletes soil of its nutrients by a certain amount. Hence management of soils either by replenishing nutrients by external sources or through proper crop management practices is necessary. In organic farming, only organic substances are allowed to feed the soils. These organic substances may be organic manure (e.g. FYM, compost), green manure (e.g. sunhemp, cowpea, dhaincha), concentrated organic manure (e.g. oil cakes, neem cakes), bio-fertilisers (e.g. Azotobactor, Rhizobium, Azolla, Blue Green Algae), vermicompost, etc.
- Complete information about the availability of organic fertilizer both inside and outside the organic farm unit help farmer to determine what organic source of nutrient or soil amendment is best for his specific fields for his specific crops in his specific situations. The rate of application of an organic fertilizer element or amendment considered should be optimum and economically feasible for one season. To determine the base rate, use information from soil analysis, existing experimental data, and local experience. Also take into consideration other soil conditions, crop to be grown, climatic conditions, irrigation possibility, management practices, economic considerations, etc.

- Taking advantage of soil management practices such as deep tillage, compaction, addition of organic matter through organic fertilizer/ compost, green manuring, inclusion of leguminous plants in crop rotation, soil productivity can be improved. In case of problem soils having excessive acidity or alkalinity, amendments like lime, gypsum, etc could be used for reclamation.

Seed and seedlings: Information on availability of organically grown seeds and seedlings is required. Seed or seedlings must be produced according to organic standards. Species and varieties cultivated in organic agriculture systems are selected for adaptability to the local soil and climatic conditions and tolerance to pests and diseases. All seeds and plant material should be certified organic.

- A wide range of crops and varieties should be grown to enhance the sustainability, self–reliance and biodiversity value of organic farms.
- Plant varieties should be selected to maintain genetic diversity.
- Organically grown varieties, and varieties known to be suited to organic cultivation should be preferred.
- Farmer should use organically bred varieties.

Organic varieties should be obtained by an organic plant breeding program. Organic plant breeding and variety development is sustainable, enhances genetic diversity and relies on natural reproductive ability. Organic plant breeding is a holistic approach that respects natural crossing barriers and is based on fertile plants that can establish a viable relationship with the living soil. Organic seed and plant materials should be propagated under organic management for at least one generation, in the case of annuals; and for perennials, two growing periods or 12 months, whichever is the longer, before being certified as organic seed and plant material.

Non-organic seeds or perennial plants (planting stock) must be managed organically for at least one year prior to harvest of crop. List all seeds/seedlings planned for use in the current season, including seeds planted previously on proposed organic fields. If farmer are using conventional treated planting materials than a note from suppliers regarding its non-availability in organic status is required.

Diversity in crop production: To maintain the diversity there should be enough greenery hedge, shrubs and trees around the farmland for sheltering birds and natural enemies, which maximise the pest and microbial activity in the soil and enhance the growth. Diversity in crop production is achieved by a combination of -

- a diverse and versatile crop rotation that includes green manure, legumes and deep rooting plants.

- appropriate coverage of the soil with diverse plant species for as much of the area as possible.
- proper management from insects, weeds, diseases and other pests, while maintaining or increasing soil organic matter, fertility, microbial activity and general soil health.

Pest, disease and weed management: Organic farming systems apply biological and cultural means to prevent unacceptable losses from pests, diseases and weeds. They use crops and varieties that are well-adapted to the environment and a balanced fertility program to maintain fertile soils with high biological activity, locally adapted rotations, companion planting, green manures, and other recognized organic practices as described in the standards. Growth and development should take place in a natural manner.

Protection of natural enemies of pests through provision of favourable habitat, such as hedges, nesting sites and ecological buffer zones that maintain the original vegetation to house pest predators diversified ecosystems. These will vary between geographical locations. For example, buffer zones to counteract erosion, agro-forestry, rotating crops, intercropping etc.

Pests, diseases and weeds should be managed by the knowledgeable application of one, or a combination, of the following measures:

- choice of appropriate species and varieties
- appropriate rotation programs
- mechanical cultivation
- thermal weeding
- natural enemies including release of predators and parasites
- acceptable biodynamic preparations from stone meal, farmyard manure or plants
- mulching and mowing
- mechanical controls such as traps, barriers, light and sound

Protection from contamination: All relevant measures should be taken to ensure that organic soil and food is protected from contamination. Farmer should take reasonable measures to identify and avoid potential contamination. Accumulation of heavy metals and other pollutants should be limited and the appropriate remedial measures should be implemented where possible. Contamination that results from circumstances beyond the control of the operation does not necessarily alter the organic status of the operation.

Water management: Efficient management of irrigation water signifies the utilization of every unit of water economically to obtain maximum returns. The goal of water management practices is to give timely irrigation with adequate quantity of water. When water is to be used for crop irrigation purposes, following five factors should be considered:

- The total salt content and chemical composition of the water
- The climate (rainfall and temperature)
- The prevalent soils and drainage conditions
- The principle crops to be irrigated
- Crop cultural practices, mainly irrigation method

A source of water may be suitable or unsuitable for irrigation after it has been examined in the light of above five factors.

3.1.3 Processing and packaging of organic products

Organic products should be protected from co-mingling with non-organic products. All products should be adequately identified through the whole process. Organic and non-organic products should not be stored and transported together except when labelled or physically separated. Certification programme should regulate the means and measures to be allowed or recommended for decontamination, cleaning or disinfections of all facilities where organic products are kept, handled, processed or stored.

Besides storage at ambient temperature, the special storage conditions such as controlled atmosphere, cooling, freezing, drying, humidity regulation, ethylene gas etc. are permitted for agricultural produce.

Pest and disease control: Pests should be avoided by good manufacturing practices. This includes general cleanliness and hygiene. Treatments with pest regulating agents must thus be regarded as the last resort. Recommended treatments are physical barriers sound, ultra-sound, light, and UV-light, traps, temperature control, controlled atmosphere and diatomaceous earth. A plan for pest prevention and pest control should be developed. There should never be direct or indirect contact between organic products and prohibited substances. (e.g. pesticides). Persistent or carcinogenic pesticides and disinfectants are not permitted. The certification programme should set up rules to determine which protection agents and disinfectants may be used.

Ingredients, additives and processing aids: Ingredients of agriculture should be 100% of organic origin. For the production of enzymes and other microbiological products the medium should be composed of organic ingredients. Water and salt should be used in organic products.

Minerals (including trace elements), vitamins and similar isolated ingredients should not be used. Preparations of microorganisms and enzymes commonly used in food processing may be used, with the exception of genetically engineered microorganisms and their products. The use of additives and processing aids should be restricted.

Processing methods: Processing methods should be based on mechanised, physical and biological processes. The vital quality of an organic ingredient should be maintained throughout each step of its processing. Processing methods should be chosen to limit the number and quantity of additives and processing aids. The approved processes are- mechanical and physical, biological, smoking, extraction, precipitation, filtration etc.

Extraction should only take place with water, ethanol, plant and animal oils, vinegar, carbon dioxide, nitrogen or carboxylic acids. These should be of food grade quality, appropriate for the purpose. Irradiation is not allowed. Filtration substances should not be made of asbestos nor may they be permitted with substances which may negatively affect the product.

Packaging: The use of material for packaging should be eco-friendly. Unnecessary packaging materials should be avoided. Recycling and reusable systems should be used wherever possible. Biodegradable packaging materials should be used. Material used for packaging must not contaminate food.

3.1.4 Organic animal production

Agriculture and animal husbandry always goes in hand in both mutual and beneficial co-existing as well as symbiotic ways. Organic livestock production intends to use native breeds of cattle, buffalo, sheep, goats and poultry. The requirements of organic production of animals are discussed below:

Breeding: Landless animal husbandry systems are not permitted. The health and performance of animals is to be promoted by appropriate rearing, the choice of suitable races and the breeding methods adopted. Animals should be bred which fit in well with the ecological limits of the various requirements and conditions to be found on an organic farm. The target is a high longevity performance by the animals. Genetic manipulation, embryo transfer and synchronisation of oestrus cycles are forbidden. Artificial insemination is allowed, but all other forms of artificial or embryo transfers, spermasexing, etc. are forbidden.

Provenance of animals: The livestock, not used for food production and except for male breeding animals, must principally come from approved organic farm operations.

Embryo Transfer: No animals deriving from embryo transfer must be kept on the farm. Bulls resulting directly from embryo transfer (ET) and their sperm are not to be used. As an interim measure, a maximum of 50% of the male breeding stock or their semen may derive directly from ET. On organic farms, natural reproduction techniques should be preferred whenever possible. In selecting animals, particular attention should be paid to the longevity performance of their forbears.

Housing and welfare: The specific ethological needs of all species of domestic animals must be taken into account with the provision of appropriate housing and the opportunity for movement and activity. The 'National Animal Protection Law' must be observed in entirety. Calves being reared and fattened are to be kept in free groups, on litter, with access to the outdoor area, from their third week on to four months of age. Keeping the animals in herds is to be preferred to keeping them separated. A journal must be kept on the grazing and free-range periods of all animals.

Stables: Stable floors with fissures or holes are forbidden. The bedding areas for all animals must be covered with straw or be dry and well insulated. Daylight must be provided. Materials and paints in stables must be innocuous (non-toxic). As far as possible, the cleansing agents and disinfectants being used must be innocuous and biodegradable.

Feed: Animals are to be fed correctly according to species (according to their natural feeding behaviour). The feeding of animals should not be in direct competition with the growing of food for humans. The basis of animal nutrition is the feed produced on the farm itself. At least 50 per cent of the feed must be produced on the individual organic farm. Conversion-fodder produced on the farm may be used in a proportion of maximum 50% of the fodder surface (conversion-farms: up to 100%). Purchased fodder serves only as supplement to this home-produced feed and fodder and should be organically produced as far as possible. The proportion of conversion-fodder should not exceed 30% of the ration of each specific livestock category.

Non-organic fodder: The proportion of non-organic fodder contained in the total fodder ration must not exceed the following percentages:

- Ruminants: 10% of the total dry matter consumption
- Non-ruminants: 15% of the total dry matter consumption per animal category

In case of losses of fodder yield, in particular because of extraordinary weather conditions, the certification body can authorize a higher percentage of non-organic fodder for directly affected livestock keepers, in a specific area and for a limited time.

Specific feeding regulations: Young mammals must be fed on the basis of natural milk, preferably mother's milk. All mammals must be fed with natural milk during a minimum period that is defined according to each specific animal species. For cattle it amounts to three months, for goats and sheep to 35 days and for pigs to 40 days. Ruminants must receive at least 60% of dry fodder matter as fresh, dried or ensilaged roughage. As for poultry, the fodder administered during the fattening period must consist of at least 65% grains and grain legumes (their products and by-products) and oil seeds (their products and by-products).

Forbidden fodder: The fodder components must be of natural condition, and the processing methods must be as natural and as energy saving as possible. Fodder must not contain any traces of genetically modified organisms. In addition following are forbidden:

- chemical-synthetic additives (urea, antimicrobial fattening promoters, enzymes, synthetic amino acids etc.)
- therapeutic chemical medication used prophylactically (sulphonamides and others), antibiotics, hormones, coccidiostatic preparations etc. Excepted are worming cures in areas of high risk (alpine meadows, permanent meadows)
- Fattening methods with forced feeding and the keeping and housing of animals in conditions that could lead to anaemia.

The addition of animal proteins, animal fats, patented preparations of fats and proteins, propylene glycol, propionic acid and any similar substances and additives that do not suit the digestion of ruminants to their fodder is forbidden. Mixtures of mineral substances or trace elements as well as vitamin preparations are allowed to meet the demand only. Natural products are recommended.

Conversion periods: In order to be regarded as organic animals, livestock not coming from organic farm operations that are purchased after the beginning of the conversion to organic farming must be kept in accordance with the rules of these standards for at least:

- 12 months as for cattle used for meat production; in any case at least during three quarters of their life
- 6 months as for small ruminants and pigs
- 6 months as for milk producing animals
- 10 weeks as for poultry for meat production, which have been put in the chicken-coop before reaching the age of 3days
- 6 weeks as for poultry for egg production

Purchase of non-organic animals (brought-in animals): About 80% of the animals must derive from certified organic farms, which means that they must have spent at least the previous twelve months on an organic farm. Purchase of animals should be from certified organic farms whenever possible. However, breeding stock may be brought in from conventional farms with a yearly maximum of 10% of the adult animals of the same species on the farm.

For the development of a new livestock, if there are not enough animals available from organic farms, they may be brought as per following age:

- laying hens up to an age of 18 weeks.
- broilers must be put into the chicken-coop on the 3rd day at the latest.
- buffalo from the moment of their weaning, but at the latest at the age of 6 months.
- calves up to 4 weeks old which have received colostrums and are fed a diet consisting mainly of full milk.
- piglets up to 6 weeks from the moment of their weaning; they must not be heavier than 25 kg.
- Two week old for any other poultry.

Newly purchased animals must be kept on the farm in order to be regarded as animals from organic production, is as follows:

- Four months for pigs
- Three months for milk producing animals

Certification programmes should set time limits (not exceeding 5 years) for implementation of certified organic animals from conception for each type of animal.

In order to develop a stock, calves and small ruminants for the production of meat may be marketed as animals from organic farming, if they are kept in accordance with the rules of the standards during the following periods before they are slaughtered:

- Calves for at least 6 months
- Small ruminants for at least 2 months.

If the quantity of animals available to supplement the natural increase or renewal of the stock, the livestock owner may, in agreement with the certification body, put in stable yearly female young cattle (animals that have not yet thrown), in a quantity coming up to 10% of the grown up cattle, as well as up to 20% of the stock of grown up pigs, sheep or goats.

Animal stocking density: The number of animals must be appropriate for the farmed area, the site and the climatic conditions. In lowland (plains), the stocking rate may not exceed 2.5 large animal equivalents per hectare of agricultural area. In mountain areas under poor conditions, the stocking rate is to be reduced. Animal stocking density is related to livestock units. To determine the appropriate density of livestock, the livestock units (dung plus urine) equivalent to 170 kg/ha/yr of agricultural area used for the various categories of animal should be set out.

Adjustments should be made for animals which produce different amounts of dung depending on their strain. If the animals are not kept year round or if they are to be allocated differently because of their age or the purpose to which they are put, then the above figures will be calculated on the average of the animals kept annually.

Animal health: Injured or sick animals must be treated. Natural remedies and healing methods have first priority, if experience shows that they have a therapeutical effect on the respective animal species or the disease that is to be treated. Chemical-synthetic allopathic treatments may be carried out on the veterinarian's prescription, if the disease or injury cannot be efficiently treated with alternative methods.

Withdrawal period: The withdrawal period between the last administering of a chemical-synthetic allopathic veterinary medicine and the marketing of foodstuff from such an animal amounts principally to the double of the legally stipulated time indicated on the package. Excepted are drugs for the draining of cows with udder problems. Before the use of draining agents, a bacteriological analysis of the milk must be carried out.

Prophylatic treatment: Prophylactic treatments with chemical synthetic allopathic medicines, antibiotics and hormones are forbidden. The uses of coczidiostatics, prophylactic iron-injections on pigs as well as the use of hormones or similar substances to control the reproduction (e.g. induction or synchronization of the heat) or for other aims are not permitted. Hormones may yet be administered to an individual animal in case of a therapeutic treatment by a veterinarian. Synthetic chemicals as worming agents and vaccinations are allowed on prescription by a veterinarian. Treated animals must be identifiable any time and unmistakably as such.

Number of treatments: If an animal is treated more than three times within a calendar year with chemical-synthetic allopathic veterinarian medicines or antibiotics (or more than once, if the productive life-cycles of less than a year), the respective animals or products of these animals must not be sold as organic. Zootechnical measures must be kept to a minimum. De-horning of young animals should only be carried out as a precaution for the safety of people and other animals and under local stunning. Permitted zootechnical measures must be carried out by qualified personnel. Castration is permitted. It may only be performed under anaesthetic and by a vet. Castration without anaesthetic of up to 14 days old piglets, lambs, goat kids and calves is tolerated.

Forbidden measures: Particularly forbidden without a vet's prescription are:

- tail-docking, except in the case of sheep when diarrhoea caused by their feed leaves no alternative and shearing does not help
- blunting of teeth
- clipping of beaks and claws in poultry
- trimming of wing-bones
- castration by means of elastic rings later than the third week of life for sheep and goats and the fifth week for calves
- caponising of roosters

3.1.4.1 Livestock species specific requirements

Cattle: Cattle are primarily to be fed with roughage. The appropriate structural balance in the daily feed ration must be observed (hay, straw, grain-whole-plant silage). Concentrates may only be used as a supplement. In summer, green feed must be provided. Exclusive year-round silage feeding is not permitted. Calves should not be taken off milk before they are 3 months old. Powdered milk substitute or powdered milk must not be given. Pure milk fattening without feeding roughage is not permitted.

Sheep: Sheep must be kept in flocks on pastureland, or in a free stable where they can move around freely and have access to the open air. Keeping them individually in boxes is only allowed for a maximum of seven days at lambing time and if they fall ill. Rams can be kept individually. During the season of vegetative growth, sheep must be put into the pastureland every day. In bad weather, it is sufficient to let them out in an outdoor run every day. In winter, all animals must have exercise area outside at least 13 times a month. Castration of males is allowed up to the age of 2 months, but the use of rubber rings only up to 14 days. Sheep should be primarily fed with roughage. The feeding Lambs are basically to be reared and fattened on mother's milk for at least 45 days. Pure milk fattening without feeding roughage is not permitted. The use of milk powder is forbidden.

Veterinary medicine: The keeping of sheep should be optimised so that worming with synthetic chemical medicaments is needed as little as possible. These medicaments are permitted by order of a vet. Preference should be given to individual treatment for claw diseases (cutting, disinfections). In bathing the feet, copper salts and formalin are to be used with discretion.

Goats: Goats must be taken to graze every day. During the birth period, mother animals must be able to move about freely for at least a day. Keeping goats individually in boxes is only permitted for a maximum of seven days after the lambs are born and in cases of illness. Male goats may be kept individually. Hormonal synchronisation of oestrus is forbidden. Goats are to be fed primarily with roughage produced on the farm. The feeding Lambs are basically to be reared and fattened on mother's milk for at least 45 days. Pure milk fattening without feeding roughage is not permitted. The proportion of concentrates (non-organically or organically grown) may only comprise 10% of the total consumption. The rearing of goats should be optimised so that worming with synthetic chemical medicaments is needed as little as possible. These medicaments are permitted by order of a vet. The use of persistent antibiotics to dry of lactation is forbidden.

Pigs: All pigs must have daily access to an outdoor-run from the 24th day of life. Farrowing sows are accepted from this in the first 24 days after farrowing. Non-lactating sows may only be shut in during feeding; otherwise they must be kept in groups, as a rule in pastures or rooting areas. Mother sows may be held individually in farrowing pens for the week before farrowing and during the suckling period. Tying up the mother sows is not allowed. The piglets may not be weaned before the age of six weeks. The pigs must have tall-growing roughage or straw available to keep them occupied. In the rearing of pigs on alpine meadows, the whey from cheese making may be fed to pigs without restriction. Pigs have to fed roughage and succulent feed to fulfil the pig's specific needs. The copper content may not exceed 10 mg per kg of the feed (on dry matter basis). Rough-age must be offered every day to breeding sows and pigs being fattened.

Poultry: The poultry must be provided with a sufficient number of places to drink and feed. On hot days, water should be offered in the runs as well.

Laying hens: If the outdoor-run is restricted, each bird must have at least 5 sqm available. The run must be covered with vegetation and have structures providing shade and protection from enemies. In addition, a completely or partially roofed area (poor weather area) with a sand bath must be available, during the whole activity period of the hens. The maximum flock size is 250 birds. If the system incorporates special buildings, it is recommended that about 3 cocks per 100 hens are kept. In the stables, there should be no more than five birds per square metre of accessible floor space. Of the total ground area, 33% must be straw-covered scratching area. With systems that have an enclosed area like a conservatory (winter garden), the conservatory area must comprise 50% of the straw-covered scratching area if it is reckoned in as part of the total ground space. The hens must have sufficient nest boxes available, containing straw or soft pile matting. The birds must have sufficient raised perches. The manure heap must be separated off and the henhouse well and naturally lit. The light period must not exceed 16 hours (other than by natural daylight in the summer). Fluorescent lighting is not allowed other than high frequency lighting. Moulting should not be artificially induced.

Laying hens in the straw-covered areas must be given a suitable grain mixture. If possible, part of the grain should be offered as whole grains in the ground bedding. Care must be taken to ensure that all birds can feed at the same time (as a precaution against cannibalism). Drinking water systems must be constructed so that birds can drink from an open water surface.

Fattening chickens: Only birds from extensively to semi-intensively reared flocks should be kept. These must be allowed to conduct their basic behavioural needs, i.e. seeking the resting area and using the raised perches provided. Birds should have no physical defects caused by breeding and no skeletal deformities. Every bird should have at least 2 sqm of outdoor-run covered with vegetation. This run must contain structures providing the hens with shade and protection from enemies. After each batch of pullets, the position of the outdoor-run must be changed to avoid a build-up of parasite populations. The outdoor-run for poultry may be used twice a year. A minimum break of 12 weeks must be observed between the usages of the same outdoor-run. A completely or partially roofed court with a sand bath (bad weather free range) must be available during the whole activity period of the hens. The maximum flock number is 500 chickens in the end phase of the fattening; and several separate flocks per farm are allowed. Fattening chicks up to 21 days may be kept in separate pens with a maximum of 1000 chicks per flock. Four flocks per pen unit are allowed. For the 4 week old chicks, access to the outdoor-run and a pasture must be permanently available. In the stables, the population density should be kept to a maximum of 20 kg live weight/sqm. The scratching area and lighting is analogous to that required by laying hens.

Turkeys: For the rearing and feeding of turkeys, the same rules apply as for laying hens and chickens. The following deviations should be noted:

- From the 7th week onward, every bird must have 10 m^2 of run available. This comes down to 9.6 m^2 for flocks larger than 125 birds, and to 9 m^2 when more than 150 birds are being reared. In the run, birds must be able to peck around. They must have a shaded area of at least 1 m^2 for every three birds.
- When penned for fattening, the population density must be no more than three birds (half males, half females) per m^2 of accessible floor space, corresponding to a maximum of 36.5 kg live weight/ m^2. In the rearing phase (1-6 weeks old) this should be reduced to 30 kg live

weight/ m^2. The ground and flooring must have enough straw to keep it dry.

- The turkey must be able to peck around in the pens as well. The buildings must be adequately lit by natural daylight. During the night, one or more dim lights must be kept on (maximum 25 Watt, red light).

If there are more than 50 birds, an inventory control card must be kept. The birds must have been on an organic farm at least since their eighth week of life.

3.2 Steps to a Successful Organic Transition

Organic farming involves following five approaches:

1. Conversion of land from conventional management to organic management
2. Management of the entire surrounding system to ensure biodiversity and sustainability of the system.
3. Crop production with the use of alternative sources of nutrients such as crop rotation, residue management, organic manures and biological inputs.
4. Management of weeds and pests by better management practices, physical and cultural means and by biological control system
5. Maintenance of livestock with organic concept and make them an integral part of the entire system

The transition from conventional to organic farming requires numerous changes. One of the biggest changes is in the mindset of the farmer. Conventional approaches often involve the use of quick-fix remedies that, unfortunately, rarely address the cause of the problem. Transitioning farmers generally spend too much time worrying about replacing synthetic input with allowable organic product instead of considering management practices based on preventative strategies. Here are a few steps new entrants should follow when making the transition to organic farming:

a) Understand the basics of organic agriculture

Since organic production systems are knowledge based, new entrants and transitional producers must become familiar with sound and sustainable agricultural practices. Transitional producers should be prepared to read appropriate information, conduct their own trials and participate in formal and informal training events. As mentioned, switching from conventional to organic farming is more than substituting synthetic materials to organic allowed materials.

Organic farming is a holistic system that relies on sound practices focused on preventative strategies. Since there are often few organic remedies available to organic producers for certain problems, prevention is the key element in organic production.

b) Identify resources

Existing organic farmers are generally very helpful in sharing valuable technical information. A good mentor should be able to provide transitional producers with knowledge, practical experience and suggest appropriate reading materials. Mentors are able to identify some of the most important challenges transitional farmers will be confronted with. Mentors may also help source production materials that are otherwise difficult to find. Producers should also contact agrologists, veterinarians and other agricultural and financial consultants, in order to learn ways to improve their current farming practices.

c) Develop transitional plan

Develop a transitional plan with clear and realistic goals. The plan should clearly identify various steps to be taken in making the transition to organic and be sure to include realistic timeframes. Identify your strengths and weaknesses. Consider ways to address any weaknesses, while building on strengths. The business side of the transitional plan should contain a multiple year budget and an effective/realistic marketing strategy. Make sure your list of expenses is comprehensive. Include all prerequisites to begin the transition; such as, mechanical weeding equipment, specialized composting equipment and applicators, additional handling equipment dedicated to the organic products, and processing equipment. Although the demand for organic products is continually growing, growers need to make sure they have a reliable market for the organic products they plan to produce.

Careful planning is very important. During the early part of the transitional period, yields are often depressed and premium prices for certified organic products are generally not yet obtainable. Use realistic yields and prices when evaluating the feasibility of your project.

In some instances, it is preferable to continue using conventional measures early on in the transitional process in order to avoid dramatic yield reduction which could jeopardize the financial well-being of the operation. Farmers who are planning to convert their livestock operation should consider certifying their fields first. This allows time to learn more about organic livestock management requirements while, at the same time, starting to produce organic feeds. Although organic certifiers generally want to see the entire farm become organic, certifiers generally allow new entrants several years of transition time before the whole farm is fully certified.

Parallel production is the simultaneous production, processing or handling of organic and non-organic crops, livestock and other products of a similar nature. Although this type of activity is highly discouraged by certifiers, some allow it, especially during the transition period. If permitted to practice parallel production, producers must be prepared to deal with significant record keeping in order to ensure traceability and organic integrity.

d) Understand soils and ways to its improvement

Since soil is the heart of the organic farming system, it is crucial that new entrants understand the various characteristics and limitations of the soils found on their farm. Soil suitability may vary significantly from one field to the next. Fields with good drainage, good level of fertility and organic matter, adequate pH, biological health, high legume content, and with less weed and pest pressure, are excellent assets. Often these fields are the first ones ready for transition and certification.

Many tools exist to assess soils. Soil chemical, physical and biological analyses, soil survey and legume composition field assessments, and field yield histories are very important and should be considered early in the transition. Unhealthy soils require particular attention.

If farmers plan to grow crops without raising any livestock, it may be necessary for them to source allowable soil amendments such as composted manure, limestone, rock dust, and supplementary sources of nitrogen, phosphorus, potassium and micro-nutrients. Even with the best of crop rotations that include green manure crops like legumes (nitrogen fixing crops), transitional growers will be challenged if they want to obtain optimal yields without additional livestock manure, compost and/or other off-farm soil inputs. When these inputs are scarce or expensive, producers may benefit from integrating livestock on their farm.

Let's not forget, under organic production, farmers must be able to recycle nutrients through proper nutrient management practices: recycling through good manure and compost utilization, crop rotations, cover crops (green manure, catch, and nitrogen fixing crops), and by reducing nutrient losses due to leaching, over-fertilization, as well as poor manure and compost management (storage, handling, and spreading).

e) Identify the crops or livestock suited for situation

Before growing a crop or raising any livestock, consider the following: degree of difficulty to grow or raise the product organically, land and soil suitability, climate suitability, level of demand for the product, marketing challenges, capital required, current prices for conventional, transitional and organic products, and profitability over additional workload.

f) Design good crop rotations

Once the crops are chosen, carefully plan the crop rotation(s) and select the most suitable cover crops (green manure, winter cover crops, catch crops, smother crops, etc.). Crop rotations are extremely important management tools in organic farming. They can interrupt pest life cycles, suppress weeds, provide and recycle fertility, and improve soil structure and tilth. Some rotational crops may also be cash crops, generating supplemental income.

On some farms, land base availability may be a limiting factor when planning your crop rotations. The transitional plan should, therefore, include crop rotation strategies. Responding to external forces such as new market opportunities may also have a significant impact on crop rotations, so farmers need to consider the effect that growing new crops has on their crop rotations and land base availability.

g) Identify pest challenges and methods of control

It is important to know the crop's most common pests, their life cycles and adequate control measures. For instance, potato beetle may be a pest of significant importance when growing potatoes; cucumber beetles in cucurbitaceous crops (cucumber, squash, and melons); flea beetle in many seedlings crops; clipper weevil and plant bug in strawberry crops.

There are several measures available to reduce pest pressure: crop rotation, variety selection, sanitation, floating row covers, catch crops, flamers, introduction of beneficial insects, bio pesticides, and inorganic pesticides. Transitional growers should be prepared to use and experiment with some of these options. When considering a new type of production, discuss pest issues with your agrologists, IPM specialists and/or other existing organic producers to optimize your chances of success.

Availability of organic supplies has improved significantly over the past few years. New pest control products containing B.t., spinosad, kaolin clay are effective and currently available to organic growers. It is often reported that the types of weeds found on the farm evolve with time as growers change the way they grow their crops and control their weeds. By keeping track of the weed population, growers will be able to refine their crop rotations and improve their control measures.

Under organic livestock management, cattlemen must provide attentive care that promotes health and meets the behavioral needs of various types of livestock. With good herd health practices, farmers rarely need to rely on conventional medicine. Organic cattlemen should, however, try to familiarize themselves with alternative remedies such as herbal/aroma therapies, homeopathy, and immune system promoters.

h) Conduct on-farm trials

Successful organic farmers continuously try new and/or innovative management practices. Practices such as cover cropping, inter-planting, and use of various soil and pest control materials need to be evaluated regularly by organic farmers. Be prepared to try new approaches.

i) Keep farm records

Record keeping is one of the most important requirements to maintain organic integrity. Farmers are expected to keep detailed production, processing and marketing information. This information includes everything that enters and exits the farm. Third party, independent inspectors require farmers to present the above mentioned documentation when inspecting the farm operation. Once the record-keeping requirements are understood and the reporting procedure established, paperwork becomes routine.

j) Avoid common mistakes

Underestimating the need for good transitional and marketing plans.

- Underestimating the need to fully understand the Organic Standard. Organic producers must understand the standard in order to know what is permitted and prohibited.
- Failing to think prevention. Transitional farmers should consider improving their crop rotation, soil and crop management skills, livestock management practices (feeding program, heard health program, grazing system, housing facilities, and husbandry).

3.3 Challenges in Organic Production

In India, many organic farmers have gained practical experience with different production methods, and suppliers of organic inputs instruct them on how to use their products. Also there is a vast heritage of traditional farming practices, which are suitable for modern organic farming.

Organic farming systems in general do not simply replace agro-chemicals with organic manures and bio-pesticides. First of all the system approach puts its focus on crop rotation, mixed cropping, recycling of biomass on the farm, ecological balance, bio-control measures and preventive methods for pest and disease management.

The existing system on conventional farm will dictate the extent to which technical challenges are a constraint to organic production. Generally the more diverse the farming system in terms of enterprises, crops, habitat and market outlet the easier it will be to produce crops organically. The range of existing crops grown is important, as it is likely that the crops farmer grow well conventionally will be his best crops organically. However, the demands of rotation may mean that farmer have to learn how to grow unfamiliar crops, such as fertility building grass/clover leys and other crops.

Soil and fertility management are definitely challenges and are perhaps not given the importance they deserve as the cornerstone of the organic system. The organic standards encourage the building of fertility through inclusion of grass/clover leys and green manures in the rotation and cycling of nutrients within the system. Grass leys are, however, not always managed well and green manures are under-utilised leaving soils vulnerable to leaching over winter. Research has shown that green waste compost can be very effective as stable source of nutrients, if a local source can be found. Maintaining good soil structure is also a challenge and the effects of compaction can manifest themselves more clearly in organically managed soils.

Weeds are often the biggest technical concern of farmers considering organic production, but the extent to which they are a problem will depend on soils and past cropping history. Investment in machinery will probably be needed but the availability of new kit such as tined and finger weeders means that weeds can be managed effectively on a field scale, especially in transplanted crops. Timing of weeding is also very crucial. The biggest problems with weeds can be in early crops where stale seedbeds are not possible and crop establishment is slower.

Rotation and diversity is the key and there are also technical solutions to many of the pest and disease problems. Many pest and disease problems on newly converted farms can be site-specific and related to pest management, so it is important to know the agriculture fields well. Although there are many permitted biocides available in the standards, the approach of substituting 'organic' inputs for conventional ones is a dubious one and should be considered only as a last resort. The risks involved with field scale production for the multiples make this approach understandable, however the organic principles of encouraging diversity and natural predators can do work. In general, organic growers find that pest and disease problems occurs less than they expected.

The additional management input required on farms should not be under-estimated and can be up to 50 per cent more. In the first two years following conversion farmers in the study have obtained crop yields from average 25-30 per cent below conventional yields and 15-20 per cent below typical yields from established organic farms. Higher prices often enable organic growers to obtain gross margins that are higher than conventional, but the economics of the whole rotation must be considered because fewer cash crops can be grown in an organic system. Research findings indicating that over-all farm income levels have the potential to be higher following organic conversion.

3.4 Commitments

Farming organically involves committing to two principles: ecological production and maintaining organic integrity. Ecological production entails using farming and ranching techniques and materials that conserve and build the soil resource, pollute little, and encourage development of a healthy diverse agro-ecosystem, which supports natural pest management. These techniques and materials include diverse crop rotations, green manuring, cover crops, livestock manure, composting, mineral-rich rock powders, etc. *Maintaining organic integrity* consists of actions that prevent contamination of organic production with prohibited materials, and that prevent the accidental mixing (commingling) of organic and conventional products. Farmers accomplish this, first of all, by not using prohibited synthetic fertilizers and pesticides; they also take precautions against pesticide drift from off-farm and other sources of contamination. Many kinds of equipment and storage areas employed in organic production must either be dedicated to organic use or properly cleaned between conventional and organic use. A considerable amount of paperwork and documentation is required to ensure organic integrity; it is one of the necessary 'burdens' of being a certified organic farmer.

Prospective organic producers should understand in advance that prohibited substances (synthetic fertilizers and pesticides) must not have been used on the land for three full years preceding harvest of the first organic crop. Farms or specific fields that do not yet meet this requirement may be considered as in **transition,** though this term does not have legal status at this time.

Organic livestock producers must make a further commitment to manage and raise their livestock in ways that are not cruel and that take account of the animal's natural behaviour. This includes providing pasture for ruminants and outdoor access for all livestock, and agreeing to restrictions on physical alterations.

Questions

1. What are the main requirements in organic production?
2. What soil information is important in organic crop production?
3. Why diversity in crop production is important?
4. What are important factors need to be considered for crop irrigation purposes under organic system.
5. Describe important steps for a successful organic transition from conventional farming?
6. Write short notes on "Challenges in organic production".
7. Write conversion periods of cattle, pig, milk producing animals, poultry for meat and egg production.
8. Fill in the blanks :
   a. Duration of the conversion period is .... months before the start of production cycle for annual crops, ... months before harvest for perennial crops and .... months for animal products.
   b. Start of conversion period is calculated from the date of .... to the certification body, when farmers commit themselves in following the standards.
   c. During conversion period, products can be labelled as "..." provided organic standards have been met for all at least 12 months.
   d. There should never be direct or indirect contact between .... products and .... substances.
   e. Landless .... systems are not permitted under organic animal production.

Chapter - 4

Biological Intensive Nutrient Management

Plant nutrients are drawn up from soil for growth and development. There are more than 90 elements present in the soil. According to the criteria of *Essentiality of Nutrients for plant growth*, proposed by *Arnon and Stout in 1952,* only 16 are considered to be essential for plant growth. Out of these, C (Carbon), H (Hydrogen), O (Oxgen), N (Nitrogen), P (Phosphorus), K (Potassium), Ca (Calcium), Mg (Magnesium) and S (Sulfur) are called *Macronutrients* (required in large quantities) and the remaining 7 elements - Fe (Iron), Mn (Manganese), Cu (Copper), Zn (Zinc), B (Boran), Mo and Cl are called *Micronutrients or trace elements* (required only in very small quantities). Among the macro elements - N, P and K are called *Fertiliser elements or primary nutrients* as they are most often applied in the form of synthetic or mineral fertilisers, and Ca, Mg and S are called *secondary elements* which are applied in smaller quantities than primary nutrients and required subsequent to primary nutrients.

The approach to plant nutrition in organic agriculture is fundamentally different from the practices of conventional agriculture. While conventional agriculture aims at providing direct nutrition to the plants by using mostly easily soluble chemical fertilisers, organic farming feeds the plants indirectly by feeding the soil organisms with organic matter. Organic soil fertility management is based on rational use of native resources achieved through crop rotation, cultivation of legumes, green manures or deep rooting plants and reutilization of organic farm by-products. Use of auxiliary resources in soil fertility management, i.e. mineral fertilisers and soil improvers that are not obtained directly from the agro-ecologic system involved and are acquired on the market, should only be employed as a second choice.

IFOAM Basic Standards as well as national regulations (NSOP) define how plant nutrition should be approached in organic agriculture and which materials are allowed, with restrictions and which are prohibited.

Following nutrition management practices are recommended:

- Biodegradable material builds the basis of the fertilization program.
- The total amount of biodegradable material brought onto the farm unit is limited.
- Animal runs should be prevented from becoming over-manured where there is a risk of pollution to rivers or groundwater.
- Brought-in material shall be in accordance with a organic standard.
- No manures containing human excrements can be used as fertiliser on vegetation for human consumption if not first sanitized.
- No chemical fertilisers containing nitrogen can be used; Chilean nitrate and all synthetic nitrogenous fertilisers, including urea, are prohibited.
- Restricted use of chemical magnesium and trace elements and/or fertilisers with unwanted substances, e.g. basic slag, rock phosphate and sewage sludge. Chemical magnesium and trace elements shall be used only after soil analysis, with prior permission of the certifier and as a supplement to organic sources.

The following products are permitted for use in manuring/soil conditioning in organic fields:

- Farmyard manure, slurry, green manures, crop residues, straw and other mulches from own farm
- Saw dust, wood shaving from untreated wood
- Calcium chloride, limestone, gypsum and chalk
- Sodium chloride
- Bacterial preparations (e.g. Biofertiliser without GMO).
- Bio-dynamic preparations (free from GMO)
- Plant preparation and botanical extracts, e.g. neem based products
- Vermicompost

The following products should be used when they are absolutely needed and taking into consideration of factors like contaminations, depletion of natural resources, nutritional imbalances, etc.

- Farm yard manure, slurry, urine, straw etc. from other farms
- Blood meal, bone meal, fish meal without preservatives
- Minerals like Basic slag, Sulphate of potash, etc

- Trace elements
- Wood ash from untreated wood
- Vermicompost from other farms

Commonly used biological nutrient management practices in organic farming are discussed below:

4.1 Organic Manures

Most plant nutrients in compost are in an organic form. Although compost is not high in nitrogen, phosphorous, or potassium, (about 2% only), these nutrients are released slowly over a period of time. Nutrients become available to plant roots at a slower rate with compost compared to inorganic fertilisers. Therefore, the nutrients are less likely to leach out. Only a fraction of the nitrogen, phosphorus and potassium applied as compost is usable by the crop during the first year and more becoming available in the years that follow.

The real benefit of adding compost to the soil lies in its ability to increase soil organic matter levels. In addition, it also improves soil fertility, buffering capacity, porosity, aeration, temperature and water holding capacity. Organic manures are prepared from organic substances obtained from different sources like plant and animal wastes of the organic farm. It helps in maintaining C:N ratio in the soil and improves the chemical, physical and biological properties of the soil. Due to increase in the biological activities, the nutrients that are in the lower depths are made available to the plants. Other advantages of organic manures are listed below:

- Excellent soil conditioner.
- Improve soil structure, decreases soil bulk density and increases infiltration rate of water.
- Increases soil buffering capacity.
- Reduces Al (aluminum) and Fe (iron) toxicity in acid soils.
- Increases carbon dioxide in plant canopy or in plant with restricted air circulation.
- It supplies growth promoting substances such as hormones.
- Acts as a source of food for soil microbes and maintains microbial activity.
- Increases soil organic, humic matter contents and overall fertility.
- Reduces soil bulk density, plants manage deeper root penetration.
- Improves the Cation Exchange Capacity (CEC), which increases

nutrient availability and reduces leaching.

- Lead to overall soil aggregate stability, which allows the soil to function at optimum levels and get optimum yield.

Organic materials are valuable by-products of farming and allied industries, derived from plant and animal sources. Organic manures is a balanced organic fertiliser and it provides both macro and micronutrients that are required by plants but in limited quantities.

There are five types of organic manures generally used in organic farming:

i) Farmyard Manure

ii) Poultry manure

iii) Oilcakes

iv) Meal group of manures

v) Compost

4.1.1 Farmyard manure (FYM)

This is traditional organic manure (cattle dung) and is most readily available to farmers. The FYM (farmyard manure, cattle manure) consists of a mixture of cattle dung, urine, bedding materials and unutilized cattle feed. These materials are collected and placed in a trench of 6 m long, 2 m wide and 1 m deep. The trench may be divided in 1 m sections. When each section is filled up to a height of about 0.5 m above ground level, the top of the heap is rounded off to the shape of a dome and plastered with earth and cow dung slurry. Five to six buckets of water may be sprinkled before plastering to ensure sufficient moisture. The plastering will save nitrogen and moisture and prevent fly nuisance. The pit should be protected from rain. Within 4 to 5 months, manure will be ready for use. The farmers' practice of storing dung and litter in exposed heaps above ground is defective and leads to loss of manurial ingredients, such as nitrogen. Such losses are prevented by the pit method.

On an average, well-rotted farmyard manure contains 0.5 per cent N, 0.25 per cent P_2O_5 and 0.5 per cent K_2O. Based on this analysis, an average dressing of 25 tonnes per hectare of FYM supplies 112 kg of N, 56 kg of P_2O_5 and 112 kg of K_2O. These quantities are not fully available to the crops in the year of application. Nitrogen is very slow acting and less than 30 per cent of it is generally available to the first crop. About 60 to 70 per cent of the phosphate and about 75 per cent of the potash become available to the immediate crop. The rest of the plant nutrients to the subsequent crop is known as 'Residual Effect'.

4.1.2 Poultry manure

This is rich organic manure, since liquid and solid excreta are excreted together resulting in no urine loss.Poultry dropping can be transformed into potent organic manure in the deep litter built up system. All that is required is to raise the birds on litter of chopped up straw, leaves or saw dust, turned over once a week. Deep litter is built up from the bacterial breakdown of litter material and droppings from the birds. Bacteria get to work in the mixture and a year later the manure is ready for use. One ton of manure can be produced by keeping 33 to 40 birds over a year time span. Ready poultry manures contain 3 % nitrogen, 2 % phosphorus and 2 % potash along with other trace elements and growth hormones. The average chemical composition of the poultry manure is as under:

Type	Moisture (%)	Nitrogen (%)	Phosphorus (%)	Potassium (%)
Fresh	75	1.47	1.15	0.48
Floor litter	24	3.03	2.63	1.40

Source : Gaur et al. (1995)

The nitrogen in poultry manure is more quickly available compared to other organic manures. Therefore, it should not be applied to the crops much in advance as we apply the FYM. If left exposed, it may lose up to 50 per cent of its nitrogen within 30 days. If the droppings come from the cages or drooping pits, super phosphate may be added @ 1 kg per day, per hundred birds. This improves the fertility quality and helps the control of flies and odour.

4.1.3 Oilcakes

Oilseeds are generally rich in manurial ingradients. After oil extraction, the oilcakes are rich in nitrogen and also contain phosphorus and potash. The nitrogen content varies from 3 to 9 per cent, depending on the nature of oilcakes. The C/N ratio is low ranging between 3 and 15 for different types of oilcakes. Due to low C/N ratio, its decomposition rate is faster than cereal and legume residues and other bulky organic manures. This nitrifies very quickly and about 60 to 80 per cent of its nitrogen is converted in available form within 2-3 months.

The manurial value of edible oilcakes like groundnut, rape-mustard, sesamum, linseed and castor and non-edible cakes such as neem, karanj, mahua, castor, etc. lies mainly in its nitrogen content. These cakes contain small quantities of P_2O_5 and K_2O.

Oilcakes need to be well-powdered before application so that they can be spread evenly and are usually decomposed by micro-organisms. They can be applied a few days before sowing or as top dressing. Mahua cake is poor in nitrogen and takes longer time to nitrify. It is better to apply mahua cake about

2 months in advance to soil before sowing of the crop.

Oilcakes on economic grounds are not recommended for cereal crops. Neem cake and castor cake is also considered to be good vermicide against some pests. The efficiency of these oilcakes can be increased by moistening it and allowing it to ferment for some time to decompose the toxic substances.

4.1.4 Meal group of manures

These are all quick acting manures suitable for all types of soil and for all crops. This group consists of blood-meal, meat meal, fish-meal, horn- and hoof-meal and bone meal. Meat meal and blood-meal are applied like oilcakes whereas fish-meal should be preferably be powdered. Horns and hooves of slaughtered or dead animals are converted into horn and hoof-meal by cooking in the bone digester, and then drying and powdering them. The average chemical composition of meal group of manures is given below:

Type	Nitrogen (%)	Phosphorus (%)	Potassium (%)
Blood meal	10-12	1.2	1
Meat meal	10.5	2.5	0.5
Fish meal	4 -10	3.9	0.3-1.5
Horn and Hoof meal	13	—	—
Raw bone meal	3-4	20-25	—
Steamed bone meal	1-2	25-30	—

Bone meal is considered useful for all soils. Best results are, however, obtained on acidic soils and soil having good drainage. It is less effective on heavy clay and calcareous soils. Bone meal is applied to the soil at sowing time or just before it. Its use as top dressing is not recommended. For vegetable and fruits, about 500 to 600 kg of bone meal per hectare is applied.

4.1.5 Compost

Compost manures are the decayed refuse like leaves, twigs, roots, stubble, crop residues etc. A large number of soil micro-organisms feed on these wastes and convert into well-rotted manure. The final product is known as compost.

Table 4.1 : Source of organic manure material

Livestock wastes	Cattle shed wastes, livestock excreta, slaughter house waste, piggery, duckery and poultry waste
Crop wastes, Tree wastes, Aquatic weeds	Cereal residues, sugarcane trash, cotton weeds, aquatic weeds, groundnut shells
Rural wastes	Bio-gas slurry
Agro-industries wastes	Press mud, paper and pulp wastes, oil cakes, paddy bran and husk, coir pith, sawdust, etc.
Marine wastes	Sea weeds, fish meal

In general, the raw materials contain cellulose, hemicelluloses, lignin, protein, minerals (ash), ether, alcohol, soluble fats, oils and waxes besides water soluble fractions (sugars, starch, amino acids, urea, ammonium salts). These components of organic wastes undergo decomposition under mesophilic and thermophilic conditions in heaps or pits and finally yield a dark coloured humified material in about 3-4 months. Some of the important sources of organic manure materials are listed in Table 4.1.

Most commonly used methods of compost production are Indore method, Banglore method and NADEP method. The methods of composting are discussed below:

Indore method: This process was developed by Howard and Wad in 1931 at Indore, Madhya Pradesh. In this method animal dung is used (as catalytic agent) along with all waste materials available on the farm to decompose in a pit having a width of 1 m and depth 1 m. Depending on the availability of waste materials, the length may be from 3 to 10 m. The materials that can be used are crop residues, cow dung, cattle shed waste, and ash. In fact, this is the first aerobic method of composting in which manure is ready within four months. Materials needed for composting may be mixed are plant residues, weeds, sugarcane leaves, grass, wood ashes, bran etc. Hard woody materials are chopped in to small pieces and are cursed. Animal dung of all the farm animals like cow, buffaloes, bullocks, horses, goats etc are collected with bedding. The urine soaked mud of the cattle shed can also be used for composting. Wood ashes reduce acidity of the compost and add potassium.

These materials are cut into small pieces and spread in layers in the pit. The first layer is of cattle shed waste filled to height of 7 to 8 cms. Then ash and cow dung are spread in layers. These layers are repeated to a height of 30 cm. It should not take more than six or seven days to fill the 3/4 length of the pit, leaving about one -fourth of the pit empty to facilitate subsequent turnings. At the end, one more layer of bedding material with wood ash and urinated mud should be provided. The sides of the pits should be sloppy and should have sufficient space between them for the carts to move freely. These materials are turned over after 15, 75 and 105 days after filling. To maintain moisture to a level of 40 to 50 per cent it is necessary to sprinkle water often in the evening and morning.

In rainy season the compost may be prepared in heaps above ground. In the areas with average rainfall, the size of the heaps may be 8 x 8 ft broader at the base and 7 x 7 ft at the top and not more than 2 ft high. In regions with heavy rainfall, compost should be prepared at safer places and protected by shed.

The Indore process is not the best method due to the relatively high labour requirement and inadequate protection from rain, sun and wind. Loss of nutrients is rapid. Upper portion of heap gradually dries as a result of poor decomposition.

Banglore method: Acharya (1939) had initiated the work of composting and this process is also called *Hot Fermentation Mechanism* of composting. It has been adopted to solve effectively the problems of safe disposal of night soil and producing high quality compost. This method overcomes many of the disadvantages of Indore Method.

This method eliminates the necessity of turning up of the materials as composting is done in trenches of 30 x 6 x 3 ft or in pits of 20 x 6 x 3 ft. Farm wastage are collected and spread in a six inch layer of cattle dung and urinated mud is followed by a layer of 1 to 2 inch thick layer of soil. This process is repeated till the heap is 1.5-2 feet above the ground level. Ultimately the heap is covered with mud plates about one inch thick. Compost will be ready within 8 to 9 months.

NADEP method: This method envisages lot of composting through minimum use of cattle dung. It requires about 90 to 120 days for obtaining the decomposed compost. In this method, tanks are constructed by using bricks or locally available materials. These tanks should be located near cattle shed or on easily approachable farm sites. The dimensions of the tanks are 3 to 4 m length, 2 m width, 1 m height with 9-inch thick wall. Lining of the bricks is made with the help of mud, which saves expenditure on cement. For circulation of air, open spaces of 15 cm width are provided at several places in the walls. About 1400 to 1500 kg of farm residues and, refuses like weeds, grasses, leaves, sugarcane trash, stubbles, stalks, roots, stems, pruning, chaff, pigeon pea stalks etc. About 90 to 100 kg (8-10 baskets) of cattle dung, 1750 kg (120 baskets) dry sieved-soil (urinated earth is more effective), and 1500 to 2000 litres of water (less during rains and in abundance during dry spells) is required for composting.

Materials are filled in the tank layer by layer. Before filling the tank, slurry made of cow dung and water should be sprinkled the floor and the walls. The first layer will be of crop residues, weeds, grass etc. (100-110 kgs) to a height of 15 cm. Over this layer biogas slurry or cow dung slurry (4-5 kg) in 125 litres water is spread in the form of thin layer. Then 4 to 5 baskets of soil are spread over this thin layer. These layers are repeated to a height of 0.6 to 0.75 m (1 and 1/2 feet above the brick level). A hut like shape may be given at the top. The whole tank is to fill within 1 or 2 days. Eleven to twelve layers are required for filling the tank to its capacity.

In case cattle dung is not available in desired quantity, collection of the same is done for 8-10 days under a shade by covering it with a light layer of soil. As an alternative practice, tank can be filled 1/3 or 1/2 of its capacity in parts. Full tank should be covered and sealed by 3-inch layer of soil (300 to 400 kg). It should be pasted with a mixture of dung and soil. Cracks should not be allowed to develop on the heaps, to check gas leakage, for that the pasting can be repeated. After 15 to 20 days the trash contracts and becomes more compact and goes down in the tank by 8-9 inches which can be refilled in the same manner and again sealed and pasted with mud and dung. In order to maintain 15 to 20% moisture, the compost is sprinkled with cattle dung and water. The microorganisms present in the material will utilise the oxygen in the air and convert the organic matter into compost.

Normally from one tank about 160-175 cubic ft of manure and 40 to 50 cubic ft of undecomposed raw material, weighing about 3 tonnes are obtained. According to an estimate, from annual collection of dung from one cow, 80 tonnes of manure can be prepared which contains 800 kg N, 560 kg P and 1040 kg K. It takes about one thousand rupees to construct a tank. Eight man-days are required for filling the pits to get it emitted and transported to the fields.

Precautions: Avoid waterlogging of compost pits and washing away of the nutrients from the already decomposed manure. If bad smell is observed during decomposition, it is due to poor aeration. Give 1-2 turnings at monthly intervals to provide more oxygen. Do not enrich more than 5 per cent P_2O_5 equivalent of rock phosphate. Higher levels of rock phosphate incorporation decrease the rate of decomposition.

The optimum value of composting parameter is given in Table 4.2.

Table 4.2 : Optimum value of composting parameter

Parameter	Optimum Value
Feed particle size	25-40 mm
Moisture content	50-60%
C/N	30/1
Temperature	55-60° C for 4-5 days
Aeration	Periodic turning (2-3 per month)

Nutrient enrichment of compost: If the organic wastes used for composting are poor in nitrogen they can be enriched by incorporating nitrogen rich organic wastes (fish meal, pongamia cake, non-edible cake, poultry manure, etc.). For phosphorus enrichment, rock phosphate equivalent to 5 per cent P_2O_5 (250 kg of Mussorie rock phosphate per tonne of organic wastes) can be mixed with 10 per cent cowdung, 5 per cent soil and 5 per cent well decomposed manure to serve as inoculum for decomposition. Phosphorus enriched compost reduces the loss of nitrogen which normally occurs during decomposition of nitrogenous rich material due to ammonia volatilisation. Experimentally, it has been found that 2 kg of phosphocompost prepared was found equivalent to 1 kg of superphosphate with respect to crop response studied.

4.2 Vermicompost

Vermicompost is a preferred balanced nutrient source for organic farming. It is eco-friendly, non-toxic, consumes low energy input for composting and is a recycled biological product. Vermicompost also has immobilised micro flora which continue their function in the soil. Every farmer can do vermicomposting in his own farm using agriculture waste.

Vermicompost is produced with the help of earthworms. Only selected species of earthworms which feed exclusively on organic matter can be used for this process of composting. The species of genus *Perionyx* are good composters, though not as efficient as *Eisenia fetida* or *E. eugeniae* which are preferred in western countries. *Perionyx species* can survive on many agricultural wastes, sericulture waste, house waste, etc. Involvement of earthworms in the composting process decreases the time of stabilisation of the waste and produces an efficient organic pool, with energy reserves as vermicompost.

4.2.1 Method of vermicomposting

Vermicomposting process involves three steps- pretreatment, composting and residue disposal. Pretreatment involves receiving refuse, shredding, decreasing its particle size and volume, separation, storage and transfer.

Vermicomposting materials: The worms feed on any biodegradable matter ranging from coir waste to kitchen garbage. All biologically degradable and decomposable organic wastes are used in vermicomposting. These include animal dung, agricultural wastes, forestry wastes, leaf litter, waste paper, cotton clothes, city garbage, bio-gas slurry and industrial wastes of biological nature. Materials with high nitrogen or proteins are best activators. High carbohydrates, if present, can lead to acidification. Ideal pH is 6.5-7.5.

Pre-treatment of composting materials: The materials collected for vermicomposting should be free from non-degradable materials like plastics, stones, glass, ceramics and metals. Agricultural wastes may be required to cut into smaller pieces for enhancing decomposition process.

The separated material should be spread in a layer of maximum of 1 foot to expose to sun for a day. To control insect pest infestation, if any, neem formulation can be used. This helps in killing some of the unwanted organisms and removes foul smell. Then the material should be mixed with cowdung slurry and cellulolytic inoculants, made a heap covered with hesian clothes or left on ground for 4/5 weeks for partial decomposition. This partially decomposed materials are now ready for vermicomposting.

Tank size: A composting tank of any convenient size can be used for vermicomposting. Preferably the pit size may be as big as 15 m x 1.5 m x 0.3 m. Alternatively, for small farmer a pit of 1.5 m wide x 1 m height will give a low cost production unit. The pit should be preferably of concrete structure and located in a place protected by shade and should be on a higher plane and free from water stagnation.

Cover of feed substrate: The pit/tank should be filled with feed mixture as follows:

- At the base of pit/tank, a layer of broken bricks is to be placed, followed by coarse sand. The thickness of layer should range between 5-7.5 cm well suited for drainage of excess water.
- At second layer, straw of paddy/wheat, banana stem peels, coconut leaves, sugarcane trash, crop stems, grass or husk should be placed. The thickness of layer should be about 30 cm well suited for aeration.
- Third layer should be of 15-30 days old cow dung with the thickness of about 20-30 cm which acts as reserve food for earthworms.
- Fourth layer or top layer should have partially decomposed waste up to thickness of 30-37.5 cm which is used for composting.
- Earthworms are introduced in between the layers @ 350 worms per m^3 of bed volume. Pit is then covered with moist gunny bag. Moist gunny bag used for reducing the moisture loss and also save worms from predators like ants.
- The beds are maintained at about 40-50% moisture content and a temperature of 20-30° C by sprinkling water over the beds. If moisture is high, dry cow dung or leaf litter should be mixed in the substrate. The pH of the substrate should be between 6.8-7.5.
- Sprinkling of water should be stopped before 3-4 days of harvesting to allow the worms to go down because of the drying of surface layers and the compost is then harvested, dried in shade and packed.

The collected worms can be released in freshly prepared beds.

- The bed should be under roof to prevent direct sun and rain.

4.2.2 Vermicompost application

Vermicompost can be applied at any stage of crops. As compared to one single dose split doses have been found to give better results. Based on several studies, all agricultural crops can be divided into following categories.

- In the first category of rainfed crop such as Sesame, Green gram, Black gram, Cowpea, Mustard, Rapeseed, etc. are considered as low nutrients requirement crops. In such crops an average of 500-750 kgs of vermicomost per hectare can give excellent results.
- In the second category of rainfed crops such as Millets (Bajra, Jawar), Castor, Isabgol, Fenugreek, Arhar, etc, farmers have been traditionally using farmyard manures. In these crops, an average of 900-1200 kgs of vermicompost per hectare can give satisfactory yields.
- In third category of crops requiring moderate irrigation such as Sunflower, Barley, Maize, Wheat, etc, farmers generally use a combination of chemical fertilisers and farmyard manures. In most of these cases an average dose of 1700-1900 kgs of vermicompost per hectare is recommended as a substitute for either of the two i.e. chemical fertiliser or farmyard manure.
- In fourth category comprising of Tobacco, Beetroot, Onion, Carrot, Sweet Potato, Okra, Coriander, Brinjal, Cucumbers, Ginger, etc, a dose of about 2400 kgs of vermicompost per hectare can be recommended as a substitute for chemical fertilisers. The use of farmyard manure can also be reduced to 50 percent level.
- The crops like Cabbage, Cauliflower, Potato, Chilli, Sugarbeet, Paddy, Tomato, Garlic, Turmeric, Broccoli, etc. use of vermicompost @ 2400-2900 kgs per hectare can be recommended to substitute half the dose of chemical fertilisers. The use of farmyard manure should be continued at previous levels.
- In high nutrients requirement crops such as Radish, Jute, Sugarcane, Banana etc, vermicompost at the rate of 1500- 2400 kgs per hectare should be applied. It has been observed that along with vermicompost and full dose of farmyard manure gives optimum yields.
- In horticulture crops especially fruit orchards depending on the stage of growth, vermicompost use of 1-20 kgs per plant is recommended.

4.2.3 Quality analysis

Chemical: The range of chemical composition of compost and vermicompost are given in Table 4.3.

Table 4.3 : Range of chemical composition of vermicompost

Parameter	Range of value
pH	: 6.0-7.5
Organic Carbon (%)	: 15 – 30
Nitrogen (%)	: 0.90 – 2.00
Phosphorus (%)	: 1.00 – 1.90
Potassium (%)	: 1.00 – 1.50
Carbon: Nitrogen	: 14-15: 1
Calcium (%)	: 2.0 – 4.0
Magnesium (%)	: 0.4 – 0.7
Sodium (%)	: 0.02 – 0.30
Sulphur (%)	: Traces to 0.40
Iron (%)	: 0.3 – 0.7
Zinc (%)	: 0.02 – 0.03
Manganese (%)	: Traces to 0.40
Copper (%)	: 0.002 – 0.010
Aluminium (%)	: Traces to 0.07
Boron (%)	: 0.003 – 0.007
Cobalt, Molybdenum	: Present in available form

Biological: Total bacter count of $2.5' x 10^6$ comprising of Azotobacter, PGPR, PSB (Phosphorus Solublising Bacteria), Actinomycetes. Also contains Gibberalline, Auxins and Cyctokinine in sufficient quantities. It contains sufficient moisture (25-35%).

4.3 Green Manures

Incorporation of leaf, stem and other plant parts in to the soil when the plant is green is called as green manuring. Green manures are the cheapest source of nitrogen for the crops. By using green manures, farmer can save on the cost of expensive concentrated organic fertilisers. Further, it will more advantageous if the green leaves are composted and then used. Green manuring, being indigenous and cost effective is an age old practice in farming system.

Use of green manure is advantageous both for crops and soil. They decompose rapidly and improve both physical and chemical properties of the soil by retaining organic matter in the soil. They provide nutrients to the crops and energy to the microbes. It can also be used as mulch to prevent soil and water losses and also control the weeds. There are different green manure crops that can be cultivated in organic farms.

The distinct benefits from green manuring are:

- Promoting microbial activities, liberating carbon di-oxide, including chemical reactions and nutrient conversion into available forms, especially the phosphates
- Soil nitrogen build-up through fixation
- Improvement of soil structure and absorptive capacity due to humus formation
- Nutrient translocation from the lower to the upper layers for plant use
- Improvement in environmental quality.

Majority of green manure crops are legumes such as Cowpea, Dhaincha, Sesbania, Sunhemp, etc. These annual green manure crops can be cultivated in all type of soils. About 6-8 tonnes of green leaf can be obtained from one hectare of land from these crops. The harvesting should be done just prior to flowering and green leaves should be incorporated into the soil. Cultivation of Dhaincha in saline and saline-alkali soils continuously for 4-5 years can reclaim the soils.

Table 4.4: Common leguminous green manure crops and nitrogen availability to the succeeding crop.

Local Name	Botanical Name	Green Matter (t ha^{-1})	Nitrogen (per cent)
Sunhemp	*Crotalaria juncia*	21.2	0.43
Dhaincha	*Sesbania aculeate*	20.0	0.43
Pillipesara	*Phaseolus trilobus*	18.3	1.10
Mungbean	*Phaseolus aureus*	8.0	0.53
Cowpea	*Vigna sinensis*	15.0	0.49
Gaur	*Cyamposis tetragonaloba*	20.0	0.34
Senji	*Melilotus alba*	28.6	0.57
Khesari	*Lathyrus sativas*	12.3	0.54
Berseem	*Trifolium alexandrium*	15.5	0.43

Similarly other perennial crops like *Triposia* and *Indigofera* can be cultivated for green leaf purpose. Bushy plants like *Glyricidia*, hedgerows (*Leucaena leucocephala, Leucaena liliricidia*) can be cultivated on bunds and vacant lands. Green leaf can be harvested one year after plantation. During the cropping season, bushes are lopped at about 0.5 m height. These lopping are used as mulch to reduce moisture loss and improve the nutrient status of soil. The leaves of perennial trees like *Pongamia*, *Neem* and *Calotropis* can also be used for green manuring purposes.

Table 4.5 : Common shrubs and trees utilized for green leaf manure

Glyricidia maculate	Leucaena leucocephala
Glyricidia sepium	Leucaena glauca
Pongamia glabra	Cassia auriculata
Deria indica	Cassia accidentalis
Ipomoea cornea	Cassia pistula
Thespesia populnea	Calotrophis gigantean
Delonix regia	Azadirachta indica
Delonix elata	Vitex negundo
Trienthama portulacastrum	Lantana camara
Tephrosia purpurea	Echhornia crassipes
Tephrosia candida	Chromoleana odorata
Didonea viscosa	Salvnia molesta

Source: Abrol and Palaniappan, 1987

There are several leguminous plant species commonly found in nature which can be advantageously used as cover crops to serve as shade plants to young plantations, forage and green leaf manure crops, and as soil binders to reduce erosion and minimize the growth of obnoxious weeds. The promising species are *Calopogonium mucumoides, haseolus trilobus* (Pillipesara), *Centrosema pubscens, Macroptilium* (Siratro), *Phaseoloides* (Kudji M.), *Dolichos lab lab* (var Ignosus), etc.

4.3.1 Green manure crops in different systems

Dryland: The scope for raising green manure crop as catch crop and recycling *in situ* or *ex situ* on dryland is very limited due to inadequate soil moisture for decomposition process. Dryland are the worst hit with respect to supply of organic matter to crops due to various constraints including poor management. Regular recycling of organic matter through compost is an alternate approach to regenerate their production capability.

Wetland: Green manure crops like *Azolla* can be grown conveniently during cropping seasons. Besides, the field bunds can be effectively utilized by raising glyricidia (at 20 cm spacing) which can yield valuable green leaf manure (up to 20 kg/bush/yr). The green leaf manure can be recycled to the field directly

2-3 weeks prior to transplanting of paddy crop or it can be used for compost making.

Farm land: The tree species having compatibility, quick growth, and higher leaf biomass can be selected and planted in the garden. These species may be medium and small canopy plants like *Sesbania* and *Glyricidia* or Perennial shades tree like *Jack*, *Erythrina*, etc. The leaf litter provides support biomass and plant nutrients apart from shade and nutrient recycling from the lower layers.

4.4 Recycling of Organic Residues

Agricultural wastes like crop residues of vegetables; weeds like parthenium, water hyacinth; dry leaves of trees etc. can be recycled by converting them into compost with little effort or by way of mulch or direct incorporation in the soil.

Cereal straw and residues: On an average, cereal straw and residues on maturity contain about 0.5 per cent nitrogen, 0.6 per cent P_2O_5 and 1.5 per cent K_2O. The quantity of nutrients in legume residues is much higher than in cereal straw. Even if 50 per cent of these crop residues are utilised as animal feed, the rest should be mobilised for recycling for their plant nutrient potential and plants like mulching.

Water hyacinth: Water hyacinth is a free floating weeds plant which grows luxuriantly in ponds, lakes and water reservoirs. Water hyacinth can be used as soil mulch, green manure and compost. The plants can directly be ploughed in the soil and allowed to decompose for a month or so before sowing of a crop.

On an average, the fresh plant contains 95.5 per cent moisture, 3.5 per cent organic matter, 1.0 per cent phosphorus (P_2O_5) and 0.2 per cent potash (K_2O). This is good source of potash. It can be converted in to compost without additional source of nitrogen. Addition of small amount of soil will accelerate the process of composting of water hyacinth. It is good for crops like vegetables, potato, upland rice, maize and jute.

Forest- litter manure: The manurial value of forest litter is as good as farm compost. A portion of the surface litter should be removed in a regular manner without adversely affecting the natural regeneration of the forests.

Hedgerows: The contour-hedgerow-farming system is based on a modification of agroforestry system in which nitrogen-fixing hedgerow species like *Desmodium rensonii, Leucacephala spp., Tephrosia candida,* etc. are planted along contours with desired food crops and other useful plants in the alleys. The technology is environmentally sound and ensures adequate control of soil erosion, helps restore soil structure and fertility, conserves soil moisture and meets food, fodder, fuel and timber needs along with income generation. It can be applied easily by the upland farmers using local resources.

4.5 Biofertiliser

Biofertilisers, in strict sense, are not fertilisers which directly give nutrition to crop plants. Biofertilisers are carrier based preparations containing mainly effective strains of some microorganisms like bacteria and fungi when incorporated with seed or soil are capable of fixing atmospheric nitrogen or solubilising insoluble phosphate in soil and making them available to the crop plants.

4.5.1 Type of biofertilisers and their utility

Depending upon the activity of mobilizing different nutrients, biofertilisers are classified as:

- Nitrogen fixer
- Phosphate and potash solubiliser and mobiliser
- Compost accelerators and enricher

a) Nitrogen fixer

The nitrogen fixing organisms can be broadly be grouped as symbiotic and non-symbiotic organisms based upon their nature of association with the plant. At present, Rhizobium, Azospirillum and Azotobacter are commonly used nitrogen fixing biofertilisers.

Rhizobium: Biological nitrogen fixation by legumes (pulse crops) in association with Rhizobium is one of the most important means of converting large quantities of nitrogen in air and makes it available to plants. The bacteria infect the legume root and form root nodules (also form stem nodules in *Sesbania rostrata*) within which they reduce molecular nitrogen to ammonia which is readily utilised by the host plant in exchange of photosynthates to produce valuable proteins, vitamins and other nitrogen containing compounds.

Rhizobium-Legume association can fix 50-300 kg N per hectare in one crop season and can leave behind substantial quantity of nitrogen in soil for the succeeding crop. The quantity of nitrogen fixed varies with the kind of legume and agro-ecological conditions available. The important point is the inocultation of efficient *Rhizobium* strain specific to each crop.

Crops like groundnut, gram, cowpea, green gram, cowpea, Bengal gram and soybean can leave behind 20-30 kg nitrogen per hectare for the succeeding crop, while the crops like lucern, pasture legumes and green manuring crops like sunhemp, sesbania, etc. add about 50-60 kg nitrogen per hectare to the soil.

Rhizobium can meet more than 50 per cent of the needs of the legume crop and about 10 to 25 per cent increase in grain yield of pulses can be obtained in most of the cases. The response varies depending on soil conditions and effectiveness of native population. In order to derive greater benefit from legumes grown in rotations and as inter-crops with non-legumes, it is necessary that legume seeds are inoculated with specific *Rhizobium*.

Azotobacter: This is free living N fixing bacterium can be used as a bio-fertiliser for several non-leguminous crops including vegetables. *Azotobacter* biofertiliser can be applied to the crops like wheat, paddy, cotton, maize, barley and the vegetables. It contributes nearly 10% increase in grain yield and 12-15% increase in vegetative biomass yield. Better effects are obtained in soils having adequate organic matter.

Azospirillum: This is very important and very widely used biofertiliser. This bacteria not only fixes the atmospheric nitrogen but also secretes growth promoting substances like IAA, GA, cytokinins and vitamins which have beneficial effects on seedling vigour and prolonged photo-synthetic activity. The important feature is that it will have intimate association with the roots of cereals and grasses. The wide adaptability of Azospirillum to varied temperature, soil pH and wide host range make this as a widely used biofertiliser throughout the world. The increase in the yield ranges between 10-30%.

Acetobacter: This biofertiliser is used for nitrogen mobilization in sugarcane, sweet potato and sweet sorghum. *Acetobacter diazotrophicus* is the most widely used species in preparing the biofertiliser. It has a potential of fixing as high as 300kg N/ha/year. The bacterium is an endophyte and therefore for optimum result the inoculum is applied to the freshly cut seed sets as sets dip treatment.

Blue Green Algae (BGA): BGA are photosynthetic prokaryotes capable of fixing atmospheric nitrogen particularly under waterlogged rice fields. They are known to contribute about 25 to 30 kg of N per hectare. Although BGA are widely distributed in different ecosystems, waterlogged rice fields provide the ideal conditions for their growth and multiplication. Besides nitrogen fixation, BGA are known to produce growth promoting substances and vitamin (B-12) which benefits rice plants. Algalisation improves the soil aggregation, water holding capacity of the soils and soil aeration. Algae add considerable amount of easily decomposable biomass to the soil and liberate organic acids and other compounds which have chelating properties.

Generally, 4-10 kg of dried BGA inoculum is required to inoculate one acre of paddy field. BGA increase 7-32% grain yield over their corresponding control. It secrets certain growth promoting substances and vitamins. In the main field, algal inoculum is broadcast over standing water at the rate of 10 kg per hectare, one week after transplanting rice seedlings.

Cyanobacteria (Blue Green Algae) like *Nostoc, Anabaena and Tolypothrix* can be multiplied by the farmers in plots of 20m x 1m x 22cm lined with polythene sheet and pH neutralised by adding lime. Starter culture is inoculated in the plots with water. A thick layer of algal growth which develops in 15 days is allowed to dry, flakes collected and used as inoculum. About 10 kg of algal inoculum can be obtained from a plot of 20 m^2.

Azolla: Azolla is a floating fresh-water fern and is distributed in lowland fields and fresh water bodies. It is a genus of small, delicate, heterosporous water fern harbours a nitrogen fixing algae *Anabaena azollae*, at all the stages of its growth and development. The dorsal lobes of the minute leaves of the plant have closed cavities on their ventral side, which house the N fixing cyanobiont *Anabaena azollae.* It is capable of growing in shallow water systems and under ideal conditions, it puts on prolific growth producing enormous quantities of biomass. It can contribute 100 to 150 kg N per hectare per year. It also adds organic matter and thus improves the physico-chemical properties of soil.

Azolla is grown in 1m^2 plots, 3 weeks before transplanting rice. Inoculation is done at the rate of 100 g per m^2 (1 q per ha). *Azolla* multiplies 100 folds without additional efforts under specific soil and environmental conditions. *Azolla* is added to the main field at the rate of 10 tonnes fresh material per hectare and incorporated into soil before transplanting rice. The other method includes growing *Azolla* along with standing crop of rice as dual crop. The incorporation method is better.

Phosphate solubilisers: A large number of microorganisms belonging to diverse families and genera and of heterotrophic to autotrophic in nature are known to have the ability to solubalise the fixed or insoluble phosphate present in the soil solution and make it available to plants. Out of these, *Pseudomonas striata*, *Bacillus polymyxa*, *Aspergillus awamorii*, *A. niger* and *Penicillium digitatum* have been known to solubalise unavailable form of P to available form. The solublisation effect is generally due to the production of organic acids by these organisms. They are also known to produce amino acids, vitamins and growth promoting substances like IAA and GA which help in better growth of plants.

Most of tropical and sub-tropical soils are deficient in phosphorus due to its fixation by calcium, iron or aluminium and no natural channels exist for the return of large annual net losses. Hence the supply is inevitably shrinking and the deposits are limited. In this context, the release of insoluble phosphates in soils and fixed phosphorus in the clay minerals by microorganisms assume significance.

This biofertiliser can be applied in the crops like oat, wheat, rice, potato, gram, lentil, mustard, chickpea, barley, soybean, groundnut, sugar beet, cabbage and tomato. Seed or soil inoculations with phosphate solubalising organisms are known to increase the growth and yield to the extent of 10 to 20 per cent. The P solubalising bacteria can be mass multiplied and mixed with the carrier material used.

Phosphate mobilising mycorrhizal fungi: There are certain fungi, which form symbiotic association with the roots of plants and help in the uptake of phosphorus. Ectomycorrhizal fungi like *Pisolithus, Laccaria, Amanita, Scleroderma, Russula, Tricholomu* form association with tree species belonging to family *Pinaceae*, *Betulacea* and *Fagceae*. These fungi increase the surface area of absorption of the roots and thus help in absorption of nutrients, especially those which are less mobile in soil solution like P. They also help in the uptake of water and protect roots from root pathogens.

Vesiculur arbuscular mycorrhizal (VAM): It is an endo-symbiont fungi, ubiquitous in nature colonizes roots of several crop plants, important in agriculture, horticulture and tropical forestry. They form appresorium inside the intercellular spaccs in the roots and other end of the hyphae extended towards soil system and thereby covering more soil spaces available for nutrient absorption. They are zygomycetous fungi belonging to the genera *Gigaspora, Glomus, Acaulospora, Enthrphopora, Sclerocystis,* etc. These fungi extract plant essential nutrients from the soil and make it available to the roots. They help plant growth through efficient absorption of phosphorus from large volume of soil through their profusely ramified mycelia, in addition to absorption of water and some micronutrients and protect the roots against pathogens.

These fungi are obligate symboints and can not be cultured on laboratory media. They can be multiplied as pot cultures, using a suitable host and the root pieces along with the substrate is used as inoculum. They are currently recommended for use in transplanted and nursery raised crops. They improve seedling growth and vigour of many plants important in agriculture and horticulture like tomato, chillies, citrus, cardamom, mango, finger millet, and forestry like *Leucaena, Dalbergia, Acacia, Casurina, Tectonneea,* etc. VAM fungi not only improves plant yield but also saves P fertilisers to the tune of 25 to 30 per cent. Combined inoculation of *Rhizobium* and VAM could meet more than 80 per cent of the N requirement of pulse crops.

Potash mobilising biofertiliser: *Frateuria aurentia* is a free-living bacterium which helps in mobilising potassium in soil. It can tolerate a wide range of pH (3.5-8.5) conditions. The bacterium solublises the potassium found in the crystalline structure of primary minerals in soil. It can fulfil up to 50 to 60 per cent of potassium requirement of the crop. The bacterium has been isolated from banana plants and is recommended for use in other crop plants. The organism has been reported to increase maximum level of potassium availability to the magnitude of 5-60 kg K per hectare.

4.5.2 Methods of placement of biofertilisers

Biofertilisers are usually inoculated by seed pelleting or seedling dipping or through soil applications depending on the nature of the crop. Although, the package of practice differs from crop to crop, however, the biofertiliser could be used in any of the methods.

Seed treatment: One packet of 200 g biofertiliser is mixed in 500 ml of water to prepare slurry and the slurry is mixed with about 10-12 kg of seeds and is air dried under the shade. Soon after the drying, the biofertiliser coated seeds are sown.

The *Rhizobium* biofertilisers are usually inoculated as seed pelleting. When the seed germinates, the root coming from the seed coat gets infected and later, the nodules are formed in the legume which fixes atmospheric nitrogen in the symbiotic process.

Seedling dipping: In this method 1 kg of biofertiliser, which is sufficient for seedling needed for one acre, is mixed with 5 litres of water in a bucket and mixed thoroughly. The roots of seedling bundles are dipped in the biofertiliser suspension for about 30 minutes and taken out. The seedlings root coated with biofertilisers are transplanted immediately. Usually *Azotobacter* and *Azospirillum* are used through seedling dipping for transplanted crops.

Soil application: In soil application, 4 to 5 kg of desired biofertiliser is mixed with 100 kg of well decomposed FYM and mixed thoroughly, so as to get a uniform mixture.

Biofertiliser mixed FYM is applied to the soil after 15 to 20 days of seedling transplantation of vegetable crops or one month after broad casting of seeds or FYM is applied in the pits before transplantation or in dragged ring, so as to ensure the placement near the root zone of perennial crop. Soon after biofertiliser is applied, it should be covered with soil, so as to avoid exposure to direct sunlight.

Phosphorus solublising bacteria (PSB) are generally used as soil application. After, the biofertiliser is applied it is covered immediately by soil under moderate moisture. The efficiency when applied to soils is limited by several factors. Most important of them are- drought and high summer temperature, water logging, unfavourable soil pH, antagonism from other organisms and nutrient deficiency.

Precautions

- Ensure that each purchased packet of biofertiliser is provided with necessary information like name of the product, name of the crop for which intended, date of manufacture, date of expiry and instruction for use.

- Use the packet before expiry on the specified crop.
- Store the biofertiliser packets in cool and dry place away from direct sun and heat.
- Never mix biofertiliser treated seeds with other organic or chemical fertilisers.
- Biofertiliser must NOT carry Genetic Modified Organism (GMO).

4.5.3 Biofertilisers and beneficiary crops

The biofertilisers are mostly crop specific which are dependent on the compatibility of the microorganisms with the beneficiary crop for delivering maximum utility.

Biofertiliser	Benefits	Beneficiary Crops
Rhizobium	10-35% yield increase, Adds 50-200 kg N/ha.	Legume crops like Groundnut, Soybean, Lentil, Red gram, Green-gram, Black-gram, Cowpea, Bengal-gram and Fodder legumes.
Azotobacter	10-15% yield increase, Adds 20-25 kg N/ha	Vegetables, Rice, Wheat, Mustard, Safflower, Niger, Sunflower, Barley, Ragi, Jowar, Fruit & Plantation Crops, Spices, Condiment, Cotton, Mulberry, Flowers.
Azospirillum	10-20% yield increase	Sugarcane, Vegetables, Maize, Pearl millet, Rice, Wheat, Fodders, Oil seeds, Fruits and Flower.
Phosphate Solubilizer	5-30% yield increase	All crops

Economics of using biofertiliser

- Saving of 40-50 kg nitrogen fertiliser per hectare.
- One tonne of *Rhizobium* inoculants is equivalent to 100 tonnes of nitrogen considering minimum fixation of 50 kg nitrogen per hectare from 0.5 kg per hectare application dose.
- One tonne of *Azotobacter* and *Azospirillum* each is equivalent to 40 tonnes of nitrogen considering maximum fixation of 20 kg nitrogen per hectare from 0.5 kg per hectare application dose.
- One tonne *Blue Green Algae* (*BGA*) is equivalent to 2 tonnes of nitrogen considering minimum fixation of 20 kg per hectare from 10 kg *BGA* per hectare application dose.

- One tonne of *Phosphate Solublisers* (*PSM*) are equivalent to 24 tonnes phosphorus pentoide considering 30 per cent solublisation at the minimum dose of 40 kg P_2O_5 per hectare (through Rock Phosphate) with 0.5 kg PSM per hectare application dose.
- Biofertiliser can also improve 10 to 40 per cent of grain yield and 15 to 30 per cent in vegetative growth.

4.5.4 Prospects of biofertiliser in organic farming

- Biofertilisers are harmless, improve the soil properties like C/N ratio, soil texture, structure, water holding capacity, etc. and maintain soil fertility.
- Secrete growth promoting substances, hormones and vitamins. They help in better nutrient uptake and increased tolerance towards drought and moisture stress.
- Confer disease resistance in crops with the help of fungistatic and antibiotic like substances.
- Solublise the insoluble phosphate and help in decomposing plant residues in soil.
- In composting processes, improve the quality and nutrient value of the compost.
- Improve seed germination and induces synchronous flowering and fruiting.
- The wasteland and lowland can be enriched and productivity may be improved by the application of biofertilisers.
- The economic benefits to cost ratio of biofertiliser is always higher so their use is acceptable in all places of cultivation.
- Biofertilisers have been found for a wide variety of crops such as cereals, legumes, millets, oilseeds, spices, vegetables, fruits, plantation crops.
- Biofertilisers are very useful for organic farming in dryland and rainfed conditions.

4.6 Soil Improvement and Amendments

Soil improvement takes place by applying following management practices in organic farming:

- Protection of the soil from strong sunlight and heavy rain by means of plant cover or mulch in order to prevent soil erosion and to preserve moisture.

- A balanced crop rotation or mixed cropping with a suitable sequence of annual crops grown on a field for preventing a depletion of the soil.
- An appropriate tillage method for obtaining a good soil structure without causing erosion and compaction.
- A good nutrient management by application of manures and permitted fertilisers according to the demands of the crops in their respective growth stages.
- Feeding and protection of soil organisms enhance the activity of beneficial soil microbes and organisms like earthworms by supplying organic material.
- To stabilize the structure, it is important to protect the soil surface with mulch or plant cover and to apply organic material (ideally compost).

4.6.1 Specific soil management practices

Nutrient release from organic matter is on going process and most plant uptake of nutrients occurs during the growth stage. Organic farmers can use a variety of management practices to simultaneously conserve nutrients and enhance soil quality. These practices include:

- Using annual soil tests to calculate appropriate amounts of organic fertilisers to add.
- Composting animal manure and other organic residues to form a more uniform and chemically stable fertiliser material.
- Using crop rotations to trap and recycle nutrients in the soil profile, increase soil tilth, and provide a diversity of crop residues.
- Applying manure and incorporating cover crops in the spring prior to plant production and nutrient uptake.
- Avoiding surface application of manure prior to rainstorms, irrigating, or when the ground is frozen or saturated.
- Using practices that enhance soil quality and reduce the potential for water runoff and wind and water erosion.
- Providing buffers or filter areas between cropping areas and water bodies to protect against nutrient and sediment movement into lakes and streams.

i) Soil testing: The soil should be tested every two years to determine the pH and available nutrients in the soil. pH is the measure of acidity or alkalinity of the soil. Too high a pH will tie up micronutrient. Too low a pH will allow the plant greater access to aluminium, restricting uptake of other necessary macro-nutrients, severely decreasing yields.

ii) Liming in acidic soil: When the pH falls below 6.0, lime is used to increase pH. There are two types of lime: Calcite and Dolomite. Calcite contains calcium. Dolomite contains both calcium and magnesium. Dolomite should be used when a soil test shows the soil to be low in magnesium. Lime also comes in pelleted and powdered form. Lime moves in the soil about one inch per year.

Equal amounts of pelleted or powered lime will raise the pH equally. Powered lime when mixed with the soil will contact the soil particles more uniformly and create a more uniform pH throughout the soil. Pelleted lime, although easier to apply will create areas of high and low pH within the soil. Lime should not be banded alone or with manure. Bands will be of very high pH and tie up micronutrient within the band rendering the manure useless in the year of application.

iii) Organic matter addition: In organic farming, soil is fed and not the crop. The soil has to hold the organic matter and create condition for the microorganisms to work on it to release the nutrient. This nutrient should be held on safely and also release as and when plants needed. Enriching and maintenance of organic matter content may be done by adopting following practices:

- Organic manures
- Composting
- Green manuring
- Organic mulching
- Crop rotation
- Mixed planting

Use of organic manure and compost: Manure, fresh or composted, makes an excellent source of nutrients for the plant. Manure should be covered immediately upon applying to the soil. As much as 30% of the nitrogen can be lost in as little as a day and a half when left exposed and not covered. Compost is easy to make and can be fortified with rock phosphate or other ingredients as needed to make a complete fertiliser.

Green manuring: Green manure with the help of nitrogen fixing bacteria takes nitrogen from the air and stores nitrogen in the plant and nodules. Legumes can be annual such as clover, dhaincha, sesbania etc. or perennial such as sweet clover, alfalfa, etc. are also used. Annuals are usually preferred to perennials. Annuals can be turned under in the fall or left for erosion protection and turned under in the spring with less chance of becoming a weed in the garden.

Organic mulching: Mulching is the process of covering the topsoil with plant material such as leaves, grass and crop residues. The kind of material used for mulching will greatly influence its effect. Material that easily decomposes will protect the soil only for short time, but will provide nutrients to the crops while decomposing. Hardy materials will decompose more slowly and therefore cover the soil for a longer time and protect against erosion. If the decomposition of the mulch material should be accelerated, animal manures may be spread on top of the mulch, thus increasing the nitrogen content.

Sources of mulching material can be the following:

- Cover crops, grass and weeds
- Crop residues (straw etc.)
- Pruning material from trees and hedges
- Wastes from agricultural processing or from forestry.

Crop rotation and mixed planting: Crop Rotation is defined as the practice of alternating the annual crops grown on a specific field in a planned pattern or sequence in successive crop years so that crops of the same species or family are not grown repeatedly without interruption on the same field. Using different crops with varying nutrient requirements assist in breaking weed cycles and in building soil fertility.

Ideally an organic crop rotation system would contain crops from different plant groups, that are seeded at different times and that have different nutrient demands. An organic field crop rotation plan should include row crops, legumes/sod crops and small grains. Production of annual vegetable crops should also include crop rotations.

Crops should be grouped and a three to four year rotation within the garden developed. Crop grouping should include members of the same family and other crops that are attacked by the same pathogen.

For example:

- Group 1: (Brassica) cabbage, cauliflower, broccoli
- Group 2: (Legumes) beans, peas, gram
- Group 3: (Root crops) carrots, beets, turnips
- Group 4: (Solanaceae) potato, eggplant, tomato
- Group 5: (Gramineae) corn, millet

A typical rotation would be:

1st yr.	Buckwheat	Legume	Group 1 2 3 4 5
2nd yr.	Group 1 2 3 4 5	Legume	Buckwheat
3rd yr.	Legume	Buckwheat	Group 5 4 1 2 3
4th yr.	Group 5 4 1 2 3	Buckwheat	Legume

Different crop plants are having different rooting systems. Some crop plant's roots spread both in depth and in lateral areas, other's root may be straight deep or shallow, spreading in all directions, and others with roots crowding in certain zones in the soil. These differences in rooting system lead to the absorption of essential nutrients from different depths in the soil. Thus, impoverishment of particular depths, where as growing crops with different root system in alternate years will lead to effective utilization of the plant nutrients in the soil. Alternating shallow with a deep rooted crop may be regarded as a method of soil restoration, because the soil nutrients are lifted from lower layer to upper layer by the action of roots.

iv) Tillage: Careful soil cultivation can improve the soil's capacity to retain water, its aeration, capacity of infiltration, warming up, evaporation etc. But soil cultivation can also harm the soil fertility as it accelerates erosion and the decomposition of humus. Depending on the cropping system and the soil type, appropriate soil cultivation patterns must be developed.

Objectives of soil cultivation:

- Loosen the soil to facilitate the penetration of plant roots.
- Improve the aeration (nitrogen and oxygen from the air).
- Increase infiltration of water.
- Reduce evaporation.
- Encourage the activity of the soil organisms.
- Destroy or control weeds and soil pests.
- Incorporate crop residues and manures into the soil.
- Prepare the site for seeds and seedlings.
- Repair soil compaction caused by previous activities.

v) Soil compaction: If soils are cultivated in wet conditions or burdened with heavy machinery, there is a risk of soil compaction which results in suppressed root growth, reduced aeration and water logging. Where soil compaction is a potential problem, farmers should be aware of the following aspects:

- The risk of compaction is highest when the soil structure is disturbed in wet conditions.

- Do not drive vehicles on your land soon after rains.
- Ploughing of wet soils can lead to a smearing of the plough sole.
- Soils rich in sand are less prone to soil compaction than soils rich in clay.
- High content of soil organic matter reduces the risk of soil compaction.
- It is very difficult to restore good soil structure once soil compaction has taken place.
- Deep tillage in dry conditions and the cultivation of deep-rooted plants can help to repair soil compaction.

vi) Soil erosion control: During the dry season, ground vegetation usually becomes scarce and thin, leaving the soil uncovered. As a result, when the rains arrive, large amounts of valuable topsoil can be washed away, leaving the land uneven with gullies and low fertility soil. Not only steep slopes but plain fields are also prone to soil erosion, and can be severely affected. Besides rain, excessive irrigation can also cause soil erosion.

Strategies for preventing soil erosion should ideally be combined:

- Reducing the erosive power of the rain drops by keeping the soil covered (with vegetation or mulch). Cropping systems should be designed in such a way that the soil is almost permanently covered with plant canopy.
- Improving the infiltration of the rainwater into the soil. The way to improve the infiltration is improving the soil structure.
- Reducing the speed of the water flowing down the slopes with the help of construction such as bunds, stonewalls, living barriers, trenches, terraces.

4.7 Living Fertile Farm

Organic farming is a systems approach where the farm is viewed as a living whole, in which each farm activity affects the others. Both raw and composted manures and biodynamic preparation are useful in organic crop production. An attention to balancing soil fertility, manures can supply all or most needs for purchased fertiliser, especially when combined with a whole-system fertility plan that includes crop rotation and cover cropping with nitrogen-fixing legumes.

Management is based on the farmer's own careful observations, plus the results of tests and analyses. The farmer needs to monitor nutrients in the soil via soil testing, and learn the characteristics of the manure and/or compost to be

used. The grower can then adjust the rates and select additional fertilisers and amendments accordingly. This leads to a modern approach in which traditional knowledge finds a renewal.

Because organic farming practices uses very limited external inputs, and reuses most on farm waste, it has a low impact on the environment. It provides an economical way of farming in which most of the costs are met at the time they are incurred. It thus offers a solution to conflicts between economics and the environment. A further benefit is the quality of the produce. Flavour and keeping quality of the foods, lustre and comfort of fibres provoke favourable comment from consumers and buyers.

Questions

1. What types of nutrient inputs are used in organic farming? Give three examples.
2. What do you understand by organic manures? Write four examples of organic manures.
3. Enumerate the advantages of organic manures.
4. What do you mean by composting? How compost is prepared?
5. What do you mean by vermicomposting? Enumerate the advantages of vermicomposting?
6. What are important components of a vermicompost unit?
7. What are biofertilisers and what are its different types?
8. How legumes help in nitrogen fixation and benefit other plants?
9. Describe the role of BGA in paddy cultivation.
10. Describe the role of phosphorus mobilising biofertilisers.
11. Why Vesiculur arbuscular mycorrhizal (VAM) is important for transplanted and nursery bed raised crops?
12. Describe different methods of placement of various biofertilisers?
13. Name five crops/small trees/shrubs that may be used for green manuring.
14. Define crop rotation. What are the principles of crop rotation?
15. What are the beneficial roles of crop rotation for managing soil fertility in organic farming?
16. Briefly describe important soil management practices in organic farming.

17. Differentiate between followings :
    a. Symbiotic and non-symbiotic bacteria
    b. Fertilisers and biofertilisers
    c. Compost and vermicompost
    d. Primary and secondary nutrients
18. A farmer engaged in Organic Farming will never use which of the following manures/fertilisers?
    (A) Compost
    (B) Single super phosphate (SSP)
    (C) Cow dung
    (D) Urea
    (E) Rock Phosphate
    (F) Sewage Sludge
    (G) Ammonium nitrate
    (H) Night soil
    (I) Biofertiliser
19. Fill in the blanks:
    a. Out of the 90 or more elements found in the soil, only............. are found to be essential for plant growth.
    b. Criteria of essentiality were proposed by .... and .... in the year........................
    c. The three elements N, P and K are called ....
    d. ...., ... and .... elements are together called secondary nutrients.
    e. When the pH falls below 6.0, ... is used to increase pH.
    f. .... and............... are used as a bio-fertiliser for several non-leguminous crops including vegetables.
    g. ....biofertiliser is used for nitrogen mobilization in sugarcane and sweet potato.
20. Match the following:

a) Iron	i) Secondary nutrient
b) Calcium	ii) Fertiliser nutrient
c) Zinc	iii) Macro nutrient
d) Carbon	iv) Trace element
e) Potassium	v) Micro nutrient

Chapter - 5

Integrated Pest, Disease and Weed Management

The problem of insect-pest, disease and weed is acute in case of all the crops and especially so in case of commercial crops. Pesticides, fungicides or chemicals are meant to control harmful pests such as insects, nematodes, diseases, weeds etc. However, excessive use of pesticides, fungicides and herbicides not only leave residues in soil, water and air but also have adverse effects on the non-target organisms such as pollinators, parasitoids, predators and wild animals. This has adversely affected the ecological balance resulting in pest resurgence, development of resistance in the pest species, diseases and environmental pollution.

5.1 Pest and Disease Management

Pest and disease management in organic agriculture consists of a range of activities that support each other. Most management practices are long-term activities that aim at preventing pests and diseases from affecting a crop. The general approach in organic agriculture is to deal with the causes of a problem rather than treating the symptoms.

Preventive measures for disease and pest management: Some important preventive crop protection measures are the following:

- *Selection of adapted and resistant varieties*: choose varieties that are well adapted to the local environmental conditions (temperature, nutrient supply, pests and disease pressure), as it allows them to grow healthy and makes them stronger against infections of pests and diseases.
- Selection of clean seed and planting material (certified seed and planting material from safe sources).
- *Use of suitable cropping systems*: diverse cropping systems, crop

rotation, green manuring and cover crops.

- *Use of balanced nutrient management*: moderate N-fertilization and steady growth makes a plant less vulnerable to infection.
- *Input of organic matter*: improves soil fertility, stabilizes soil structure and supplies substances that strengthen the plant's own protection mechanisms.
- Compost can reduce disease problems due to the presence of microorganisms. They either competes with pathogens for nutrients, produce antibiotics that reduce pathogen survival and growth, or parasite on the pathogens. There is also an indirect effect on crop health.
- *Application of suitable soil cultivation*: regulates weeds that serve as hosts for pests and diseases.
- *Use of good water management*: flood irrigation and water on the foliage encourages pathogen infections.
- *Conservation and promotion of natural enemies*: provide an ideal habitat for natural enemies to grow and reproduce.
- *Selection of optimum planting time and spacing*: sufficient distance between the plants allows good aeration and reduces the spread of a disease.
- *Use of proper sanitation measures*: removes infected plant parts from the ground to prevent the disease from spreading, eliminates residues of infected plants after harvesting.

Curative crop protection methods: If all preventive crop protection practices fail to sufficiently prevent economic losses to the farmer, it may be necessary to take curative action. Curative action means controlling the pest or disease once it has already infested the crop. Several options exist in organic agriculture:

- Biological control with natural predators or antagonistic microbes.
- Natural pesticides based on herbal preparations or other natural products.
- Cultural and mechanical control with traps or hand picking.

5.1.1 Use of bio-control agents

Biological Control is defined as the reduction of pest populations by natural enemies. Natural enemies of insect pests, also known as biological control agents, include **parasitoids**, **predators** and **pathogens**. Parasitoids are species whose immature stage develops on or within a single insect host, ultimately killing the host. Most have a very narrow host range. Many species of wasps and some flies are parasitoids.

Predators, such as lady beetles and lacewings, are mainly free-living species that consume a large number of preys during their lifetime.

Biological control agents of plant diseases are most often referred to as antagonists. Pathogens are disease causing organisms including bacteria, fungi, and viruses.

Biological control agents of weeds include herbivores and plant pathogens. They kill or debilitate their host and are relatively specific to certain insect groups.

5.1.1.1 Insect parasitoides

The egg-parasitoids Trichogramma: Trichogramma are minute parasitic wasps (8 to 10 adults can sit on a pin head) and use host eggs for development and multiplication. The tiny adult parasitoids search for host eggs and parasitise them i.e. they lay their own eggs within the eggs of the pests. On hatching, the parasitoids larva feed on the embryonic contents of the egg and completes its development within. The parasitised eggs turn uniformly dark in about 4 days later. Thus, instead of a caterpillar, a Trichogramma adult comes out of the host egg. A single Trichogramma, while multiplying itself, can destroy over 100 eggs of the pest. Among all parasitoids, Trichogramma is the most important and widely used for control of a variety of pests throughout the world. They are extensively used for the control of bollworm, sugarcane borers, rice stem borers, rice leaf folder, apple codling moth etc. Trichograma eliminates the pest before it has a chance to damage the crop.

Trichogramma are mass produced and supplied in the form of Trichogramma-cards, each containing about 20,000 ready-to-emerge parasitoids. Such cards are safely kept in packets. Each card may be cut into small pieces and distributed all over the field. The pieces may be stapled to the underside of the leaves or they may be placed in small boxes fitted with fine wire mesh and tied to the plant. The parasitoids emerge and disperse in search of host eggs. Alternatively, the adult parasitoids may be allowed to emerge in the packets and distributed in the field on the same day.

Release of 1,00,000 to 2,50,000 Trichogramma per hectare is recommended. More parasitoids may be released depending upon the crop and pest density. Since Trichogramma is an egg parasitoid, its releases should synchronise with the egg-laying period of the concerned pest. Trichogramma kills the pest in the egg stage itself, thus preventing potential damage to the crop by the caterpillars. The parasitoids themselves search the host and destroy them while multiplying themselves. They are self-sustaining and have no adverse effects on plants, environment or other beneficial organisms.

Larval-parasitoids, Goniozus and Bracon: *Goniozus nephantidis* and *Bracon brevicornis* are larval parasitoids of the coconut black-headed caterpillar, *Opisina arenosella*, which is a serious pest of coconut palms. These parasitoids are mass-produced and widely used for control of this notorious pest.

When the parasitoids are released in the field, they go in search of the caterpillars feeding on coconut leaves. On locating a caterpillar, the parasitoids paralyses it by stinging and lays up to 15 eggs on it. The eggs hatch in 1 or 2 days. The parasitoid larvae feed on the paralysed host and complete their development in 3-4 days. They form cocoons on or near the remains of the host. After 7-10 days, adult parasitoids emerge and go in search of fresh caterpillars. A female can lay up to 120 eggs and these are distributed on several caterpillars. Thus the caterpillars are killed while the parasitoids multiply themselves. Releases of 3000 to 4500 parasitoids per hectare/season are recommended.

***Pupal-parasitoids, Spalangia* spp:** Several parasitic wasps like *Spalangia* spp. attack the puparia of housefly and other filth flies in their natural habitat. Such parasitoids are mass-produced and released in fly breeding areas for biological control of filth flies. *Spalangia* spp. Dwell within the manure, never becoming pests themselves.

The female parasitoids are ready to mate and oviposit immediately after emergence. The parasitoids search and lay their eggs within the fly puparia. The eggs hatch and the parasitoids larvae develop by consuming the fly puparia. Finally the fly puparia are destroyed as the parasitoids complete their development. It completes its development from egg to adult emergence in 15 to 18 days. Single parasitoids lays its eggs in about 100 fly puparia in about 15 days, thus causing their death as they multiply themselves.

Spalangia spp. may be released in poultries, dairies, piggeries, garbage dumping yards etc. where the fifth flies breed. They are purely beneficial, never themselves becoming nuisance.

5.1.1.2 Insect predators

Predacious green lacewings, Chrysoperla species: *Predacious* green lacewings, *Chrysoperla* spp. is well known predators of soft-bodied sucking pests such as aphids, thrips and whiteflies. Species such as *Chrysoperla carnea* feed on the eggs and newborn larvae of Lepidopterans like cotton bollworms. The adult of *Chrysoperla carnea pale* green, possessing net-like wings. It is popularly known as green lacewing. It lays its characteristic stalked green eggs in small groups on the leaf or other parts of the plant. A female can lay 300 to 400 eggs.

The eggs of *Chrysoperla* hatch in about 4 days. The larvae (or grubs) which possess sickle-shaped mouthparts, wander on the plants in search of prey and crush them dry. The larval period lasts for about 8 days. During this period, each larva consumes several hosts every day. The larva pupates within a round. Silken cocoon on the leaf. The entire life cycle is completed in 17 to 20 days. The adult lives for 30-40 days.

Chrysoperla are mass produced in insectaries and supplied to farmers generally in the form of eggs. The eggs are de-stalked and then packed in boxes using suitable inert medium. These may be distributed on the plant. These predators should be applied twice during the cropping season with an interval of 15 days at the rate of 50,000 per hectare.

Predacious ladybird beetles: *Ladybird Beetles* are well known predators of mealy bugs, scale insects and other soft bodied sucking pests. Three species viz. *Cryptolaemus montrouzieri*, *Scymnus coccivora* and *Nephus spp* give effective control of mealy bugs in coffee, citrus, grapevine, mango etc. Two other species *Chilocorus nigrita* and *Pharoscymnus horni* destroy scale insects of sugarcane, coffee, sapota and several other crops. *Menochilus sexmaculatus* is a voracious predator of aphids.

The adult ladybird beetles, being winged, search for mealy bugs or scale insects and feed on them throughout their life. They live for about 45 days. They lay their tiny eggs in the colony of the pest. Each female can lay 100 to 250 eggs or more depending upon the species. Eggs hatch in 5 to 6 days and the grubs voraciously feed on all stages of the host insects. They are particularly fond of host egg sacs and gravid females, and often completely wipe out the pest colonies. They then enter into the pupal stage, usually in groups, in the crevices of the bark, fallen dried leaves, in leaf folds, infested plant surface, etc. the pupae do not move. Adult beetles emerge from the pupae in 8 to 10 days. They mate and again start feeding and breeding. Thus, the ladybird beetles destroy the pest while building up their own populations.

The ladybird beetles are mass-produced and made commercially available to the farmers in the form of grubs or adults. Release of 1500 to 2500 grubs per hectare at the first sign of infestation is recommended.

5.1.1.3 Antagonistic organisms

The various organisms employed, as bio-control agents are *Trichoderma* spp., *Aspergillus* niger, *Gliocladium virens*, *Bacillus subtilis*, *Pseudomonas fluorescence*, *Paecilomyces lilacinus* and *Verticillium chlamydosporium* etc. among these the most commonly employed and commercially available antagonists are discussed here.

Trichoderma: *Trichoderma harzianum* and *Trichoderma viridae* are commonly used fungal antagonists for disease management in nursery and main field. While *T. harzianum* can be used for managing fungal diseases and nematodes, *T. viridae* is specific against fungi. Hence, it is advantageous to use *T. harzianum*. The minimu requirement of the count of *Trichoderma* is 10^6 spores/1g of formulation. Many firms are manufacturing *Trichoderma* formulations and both neem and talc formulations are available. *Trichoderma* is highly effective in reducing wilt disease of pepper.

Pseudomonas florescenes: The foliar application of this insect pests like leaf minor, rice leaf folder etc. and diseases like bacterial wilt, fungal diseases and nematodes. Regular sprays are advocated to minimize the disease incidence in main field and these are believed to induce resistance in plants. Treating the nursery @ 50 g/sq. meter with this antagonistic bacteria induces disease resistance against bacterial wilt in tomato.

Bacillus thuringensis (B.t.): This is the world largest selling bio-pesticide and contributes to 80-90 per cent of bio-pesticides produced in the world. It is a pathogen of *Lepidopteran* pests and causes gut paralysis on ingestion killing the pest. Both liquid and powder formulations are commercially available. Several strains of *B.t.* are available of which *kutarki* and *israelensis* are more popular.

The caterpillars ingest *B.t.* while feeding on the treated plants. After about 24 hours of infection, their feeding activity gradually diminishes, the larvae become sluggish, turn black with signs of vomiting and diarrhoea and finally die in about 6 days. *B.t.* is used for the control of lepidopterous pests infecting cabbage and other cruciferous crops, cotton, tomato, gram, groundnut etc. It has no adverse effects on plants or environment.

Verticillium lacanii: This is a pathogen of aphids and white flies and one of the most commercially developed microbial pesticides against *Coccus vividis, Archips termias, Empoasca* sp., *Aphis gossypii* and *Tetranychus* sp.

Beauveria bassiana: *B. bassiana* has been found to be effective against *Spodoptera littura*, *Helicoverpa armigera*, *Chilo partellus*, *Scirpophaga certucas* and *Myllocerus* sp. However, care should be taken while using this organism in sericulture areas as the fungus causes muscardine in silkworms.

Paecilomyces lilacinus: *P. lilacius* is a potent egg parasite of root knot nematode *Meloidogyne incognita* and is being used in many agricultural, horticultural, plantation crops and also in mulberry cultivation.

Vesicular-arbusculur Mycorrihza (VAM): VAM increase the water holding capacity, nutritional status of the soil and also help in promoting growth and disease resistance. This can be cultured in the farmer's field itself. For this solarize the soil for 30 days, cultivate the soil and add VAM culture and grow either ragi or jowar. This facilitates growth of VAM in soil. This is added to soil for growing seedlings and to the main field before sowing or transplanting.

Nuclear Polyhedrosis Viruses (NPV): Heliothis and Spodoptera Nuclear Polyhedrosis Viruses (NPV) are baculoviruses, which are target specific and infect a number of important plant pests of the lepidoptera. They are extensively used in many countries. The most important advantage of these groups of viruses is their harmlessness to non-target organisms including natural enemies of pests. They are compatible with other biopesticides. They can be mass produced very easily.

These NPV can be applied at the rate of 250 larval equivalent (LE) per hectare on chickpea, tomato, maize, sorghum, alfalfa, sunflower, cotton, pigeon pea, etc. NPV should be mixed with water along with 0.5 per cent jaggary (local sugarcane product i.e. gur) and 0.1 per cent Ranipal (Blue) and sprayed on the infested crop preferably in the evenings. NPV should be mixed with 0.25 per cent boric or tannic acid on crops like groundnut, cole crops, beet root, etc.

The caterpillars, while feeding on the plant, ingest the virus. The virus multiplies rapidly within the body of the caterpillars and kills them within 6 days. Such diseased larvae die with their head hanging down, purify, releasing a mass of virus which is infective to fresh caterplillars. Being species-specific, NPV does not affect non-target organisms, including *Trichogramma*, *Chrysoperla* and other bio-control agents. It is safe to plants and environment.

Biological agents to control pests of different crops

Sl. No.	Biological Agents	Pest	Crop
1.	*Trichogramma brasiliensis* @ 1.0 cc/ac. once in 10 days, (Egg parasitoid)	*Lepidopteran, Heliothis sp.*	Cotton, Tomato
2.	*Trichogramma chilonis* @2 cc/ac once in 15 days	Borers	Sugarcane, paddy, pulses Vegetables
3.	Nuclear Polyhedrosis Virus (NPV) @100-200 LE/ac	*Spodoptera sp* & *Heliothis sp*	Vegetables
4.	*Chrysoperla Spp.* @5000 - 10000 eggs /ha, 3 – 4 times in 15 days, (Green lace wing)	Prudenia, Caterpillars, White flies, thrips, aphids	Vegetables
5.	*Beauveria bassiana* @ 1.0% affects the young stage, Helicoperva, Spodoptera, borers, hairy caterpillars, mites, scales, etc.	Vegetables, cereals, fruits	
6.	*Metarhizium anisopliae* @ 0.5 - 1.0 % affects all stages	White grubs, Beetle grubs,	Sugarcane, groundnut,

Contd...

		caterpillars, Semiloopers, mealy bugs, BPH	rice, potato, cotton, cereals
7.	*Verticillium lecanii* @ 0.5 - 1.0 %, affects all stages	All sucking soft bodies insects	Sugarcane, groundnut, rice, potato, cotton, cereals
8.	Phascilomycetes	Nematodes	All crops
9.	*Bacillus thuringiensis* @ 0.3 - 0.4 % affects all stages	Helicoperva, spodoptera, borers, hairy caterpillars, mites, scales, etc	Vegetables, cereals,fruits
10.	NPV -Nuclear Polyhedrosis Virus of *Spodotera litura* @250 - 500 ml/ ha 2 - 3 time at 10 days interval	*Spodotera litura*	Cotton, groundnut, pulses, cabbage, chillies
11.	NPV Nuclear Polyhedrosis Virus of *Helicoverpa armigera* @250-500 ml/ ha, 2 – 3 time at 10 days interval	*Helicoverpa armigera*	Cotton, groundnut, pulses, cabbage, chillies

5.1.2 Biopesticides

The pesticidal compounds derived from living organisms are biopesticides. These products are more selective, eco-friendly and leave no toxic residues in the environment unlike synthetic insecticides.

Limonene and Linalool: Citrus oils are extracted from oranges and other citrus fruit peels and refined to make the compounds d-limonene and linalool. Both compounds are generally regarded as safe, and are used extensively as flavourings and scents in foods, cosmetics, soaps, and perfumes.

Limonene and linalool are contact poisons (nerve toxins) and may have some fumigant activity against fleas. They have low oral and dermal toxicities. Both compounds evaporate readily from treated surfaces and have no residual. They are effective against all external pests of pets, including fleas, lice, mites, and ticks. Commercial products (usually called "d-Limonene") are available as sprays, aerosols, shampoos, and dips for pets.

Neem: Neem or neem oil is extracted from the seeds of the neem tree, *Azadirachta indica*, a native of India. The neem tree supplies at least two compounds with insecticidal activity (azadirachtin and salannin), and other unknown compounds with fungicidal activity. Azadirachtin acts as an insect feeding deterrent and growth regulator. The treated insect usually cannot moult to its next life stage and dies. It acts as a repellent when applied to a plant and does not produce a quick knockdown and kill. It has low mammalian toxicity and does not cause skin irritation in most formulations.

Neem has some systemic activity in plants. Currently registered products for ornamental pest control claim activity against a variety of sucking and chewing insects. Neem is most effective against actively growing immature insects.

Neem has been evaluated against more than 500 species of insects at global level and most of these insects have been found to be susceptible. Neem products give good control of grass hoppers, leag hoppers, plant hoppers, leaf minors, whiteflies, scales, mealy bugs, caterpillars such as diamondback moth on cole crops and gram pod borer on red gram, cotton, tomato etc. Neem kernel powder at 1 to 2 per cent gives good control of pulse beetle and other storage pests. Neem oil is used to control powdery mildew.

Neem and Pongamia cakes suppress nematodes. Soil application of neem cake (250 kg/ha) at planting and repeated at flowering reduces leaf minor and fruit borer in tomato, shoot and fruit borer in brinjal, leaf minor and fruit fly in cucurbits, pod borers in vegetable pigeon pea, dolichos lab lab, etc. Coconut mite infestation can be reduced by tying cloth bags containing neem cake in crown region. Cake should be replaced at 45 days interval.

Pyrethrum/Pyrethrins: Pyrethrins are highly concentrated active compounds which are extracted from the daisy-like flower of *Chrysanthemum cinerariaefolium*. When the flower is ground into a powder, the product is called a pyrethrum. Pyrethrum is the most widely used botanical insecticide. Synthetic insecticides that mimic the action of the pyrethrins are known as pyrethroids (e.g., bifenthrin, cyfluthrin, and permethrin).

Most insects are highly susceptible to low concentrations of pyrethrins. The toxins cause immediate knockdown or paralysis on contact, but insects often metabolise them and recover. They act specifically by disrupting the sodium and potassium ion exchange process in insect nerve fibres and interrupting the normal transmission of nerve impulses. Pyrethrins break down quickly; have a short residual, and low mammalian toxicity, making them among the safest insecticides in use. However, people may have allergic skin reactions and cats are highly susceptible to poisoning (e.g., flea powder).

Pyrethrins may be used against a broad range of pests including ants, aphids, roaches, fleas, flies, and ticks. They are available in dusts, sprays, and aerosol. Trade names include Pyrenone and Pyrellin. The use of Pyrethrum/ Pyrethrins should be done in consultation with certification body due to its restricted use in organic farming.

Rotenone: Rotenone is extracted from the roots of two tropical legumes, Lonchocarpus and Derris. It inhibits the conversion of nutrients into energy at the cellular level (cellular respiration). Insects quickly stop feeding. Death occurs several hours to a few days after exposure. Rotenone degrades rapidly when exposed to air and sunlight. It is not phytotoxic, but is extremely toxic to fish, and moderately toxic to mammals. It is more toxic to mammals by inhalation than by ingestion, and skin irritation and inflammation of mucous membranes may result from skin contact. Wear protective clothing and a mask. It may be mixed with pyrethrins or piperonyl butoxide to improve its effectiveness.

Rotenone is a broad-spectrum contact and stomach poison that is effective against leaf-feeding insects, such as aphids, certain beetles (asparagus beetle, bean leaf beetle, Colorado potato beetle, cucumber beetle, flea beetle, strawberry leaf beetle, and others) and caterpillars, as well as fleas and lice on animals. It is commonly sold as a 1% dust or a 5% powder for spraying, and trade names include Bonide Rotenone, Rotenone-Pyrethrins Flea Dip, Rotenone/Pyrethrum spray, etc. The use of this pesticide in organic farming is limited.

Ryania: Ryania is extracted from the stems of a woody South American plant, *Ryania speciosa*. Although a slow-acting stomach poison, it causes insects to stop feeding soon after ingestion. It works well in hot weather. Ryania is moderate in acute or chronic oral toxicity in mammals. It is generally not harmful to most parasites and predators, but may be toxic to certain predatory mites. Ryania has longer residual activity than most other botanicals.

It is used commercially in fruit and vegetable production against caterpillars (corn borer, corn earworm, and others) and thrips.

Sabadilla: Sabadilla comes from the ripe seeds of the tropical lily *Schoenocaulon officinale*. The alkaloids in Sabadilla affect insect nerve cells, causing loss of nerve function, paralysis, and death. The dust formulation of sabadilla is the least toxic of all registered botanical insecticides. However, pure extracts are very toxic if swallowed or absorbed through the skin and mucous membranes. It breaks down rapidly in sunlight and air, leaving no harmful residues.

Sabadilla is a broad-spectrum contact poison, but has some activity as a stomach poison. It is commonly used in organic fruit and vegetable production against squash bugs, harlequin bugs, thrips, caterpillars, leaf hoppers, and stink bugs. It is highly toxic to honeybees, however, and should only be used in the evening, after bees have returned to their hives. Formulations include baits, dusts or sprays.

Horticultural Oil. Considered effective and safe, horticultural or summer oils (e.g., petroleum oils and vegetable oils) are used to control insects and diseases. Commercial products are highly refined, and formulated as dormant or summer oils. Dormant oils are heavier oils, are more likely to damage plant tissue, and are used on dormant plants to control over wintering insects (e.g., aphids, spider mites, and scales). Summer oils are a lighter version of dormant oil and can be applied to actively growing plants to control aphids, mites, thrips, scales, mealy bugs, and their eggs. Oils coat the insects and suffocate them; therefore, thorough coverage is important.

If the product is not used properly, plant damage can occur. This may happen when too much oil is used, plants are water stressed, temperatures exceed 90°F or when dormancy is mistaken (i.e., spraying too early in the fall). Temperatures must be above 45°F for dormant oil application for proper viscosity and coverage on plants.

Insecticidal Soap. Insecticidal soaps are made from plant oils (neem, pongamia, cotton seed, olive, palm, or coconut) or animal fat (lard, fish oil), but are generally not considered botanicals. They are made from the salts of fatty acids, which are in the fats and oils of animals and plants.

Soaps are thought to physically disrupt the insect cuticle (outer skin), but additional toxic action is suspected. Soaps act on contact and must be applied directly to the insect to be effective. No residues remain on plants. They are effective against soft-bodied insects like aphids, some scales, psyllids, whiteflies, mealybugs, thrips, and spider mites. Hard-bodied insects (e.g., adult beetles or wasps) are not harmed because of their tough, chitinous bodies.

Some plants may be sensitive to soaps, resulting in leaf burn. Plants that have hairy leaves may be more susceptible to soap injury than smooth-leaved plants. Apply the soap spray on a small area of the plant to check for phytotoxicity.

The potassium soaps (soft soap) of Neem and Pongamia oils at 1 per cent are effective against diamond back moth, leafhopper, aphids, leaf miners, fruit borers and fruit fly in vegetable crops. They are also effective against mango leafhopper and citrus leaf minor. Efficacy may be reduced at high temperatures.

Biopesticides and IPM products for various crops

Crop	Pest/Diseases	Biopesticides
Cotton	Bollworms	Traps, Lures, Trichogramma, BT, NPV
	Whitefly, Jassids, Thrips	Neem 1500 ppm
	Mites	Chrysoperla, Baeuveria, Verticillium
	Wilts and Leaf spots	Trichoderma, Pseudomonas
Rice	Yellow Stem borer, Leaf folder	Traps, lures, BT, Trichogramma
	Hoppers	Neem 1500 ppm
	Sheath blight and leaf spots	Baeuveria
Pulses	Bollworms or cutworms	Traps, Lures, BT, NPV, Trichogramma
	Wilts	Trichoderma, Pseudomonas

Contd...

Tomato,	Heliothis	Traps, Lures, BT, NPV, Trichogramma
Capsicum	Mites	Trichoderma, Pseudomonas
Brinjal	Fruit borer	Traps, Lures, BT, NPV,
Okra	Mites	Trichogramma Neem 1500 ppm, verticillium

5.1.3 Cultural methods

Cultural control is generally preventive in nature and hence need to be carried out at appropriate time. Cultural control is the cheapest of all control measures. However, this technique is indirect and intangible. The following practices can be gainfully adopted in different cropping systems to suppress certain pest species.

Tillage operation: Ploughing or hoeing helps to expose stages of soil inhibiting insects to sun or to the predatory birds. Earthing up of soil in sugarcane reduces seedling borer infestation. Scraping of bunds in paddy fields destroys eggs of grasshoppers.

Field and plant sanitation: Regular removal of weeds, pest-affected parts of the plants, crop stubbles and their destruction will eliminate the sources of infestation of the pest. Destruction of bored shoots and fruits of brinjal prevents further build up of the pest population. Many virus diseases like leaf curl, bud and stem necrosis in crops like tomato, groundnut, sunflower can be minimised by uprooting the infected plants.

Crop rotation: growing of a non-host crop after a host crop of the pest will break the breeding cycle of pest species and reduce their populations. Like wise, crop rotation prevents the build up of plant pathogen in soil.

Growing pest resistant varieties: Certain varieties of crops are less damaged or less infested than others by insects. The resistant varieties have physical and physiological features, which enable them to avoid or withstand pest attacks.

Trap cropping: The growing of susceptible or preferred hosts as trap crops along with the main crop attract the pests in large numbers and these can be selectively killed. This minimise the damage on main crop. For example, marigold is grown as trap crop for tomato fruit borer and mustard as trap crop for cabbage pests. In cotton, ladies finger can be grown as a trap crop.

Plants more preferred by the pest for egg laying and feeding are grown as trap crops on the bunds of the man crop or at random.

Crops	Pest	Trap crop
Cotton, Groundnut	Spodoptera	Castor, Sunflower
Cotton, Chickpea	Helicoverpa	Marigold
Pigeonpea	Helicoverpa	Marigold
Groundnut	Spodoptera	Castor
Cotton	Spotted bollworm	Bhendi

Adjusting the time of sowing: The simultaneous sowing of crops in a locality helps in reducing pest damage. Many a times early sown crops escape pest attack. Jowar crop sown before the end of June month usually escapes attack by shoot fly.

Wider spacing: Wider spacing between rows and plants, and lesser crop density often prevent build up of populations of certain pests and also incidence of certain diseases. Wider spacing reduces incidence of brown plant hopper in rice and woolly aphid in sugarcane. In tomato, the spacing of more than 1m between rows and 75 cm between plants reduces disease incidence and its spread.

Growing of barrier crops: Maize is traditionally used by farmers around many vegetable crops. This acts as a barrier for pest and vector species.

Water management: Draining of water for a few days in paddy fields suppress brown plant hopper population. Flooding of fields wherever possible kills root grubs and termites.

5.1.4 Mechanical methods

Hand picking of egg masses, gregarious larvae and sluggish adults, and their destruction helps in reducing pest populations in certain situations. Grown larvae of *Helicoverpa* sp. and Red headed hairy caterpillar can be hand collected and destroyed. Fixing of tin bands over coconut trunks prevents damage by rats. In cabbage, scouting and mechanical destruction of gregarious early instar larvae of tobacco caterpillar and leaf webber is very effective.

Pheromones: Sex pheromones are complex organic compounds emitted by insects, mostly females to attract the opposite sex for mating. Such chemical compounds, specific to each species, can be synthesized in the laboratory and made available as lures for use in traps. These pheromones attract the insects in large numbers to the traps where they get trapped and killed. Female sex pheromones are available for all these species of cotton bollworms like *Helicoverpa armigera, Sapodoptera litura, Pectinophora gossypiella, Earias vittella* and *Earias insulana* as also for rice stem borer, diamondback moth, apple codling moth etc.

Sex pheromones are helpful in the early detection of pests and their timely control. Mass trapping of males results in depriving the females of mating opportunities, thereby affecting reproduction. Pheromone is non-toxic and species specific. It is safe to other organisms, plants and environment.

To manage brinjal shoot and fruit borer, 50-60 pheromone traps per hectare can be erected. Lures like methyl eugenol, cue-lure can be used to suppress fruit flies. Red palm weevil on coconut can be controlled by using sex pheromone traps.

Ready-to-use sex pheromone lures and traps are commercially available. These may be installed in the field at 30 to 60 cms above the crop canopy. Generally, at least 5 traps are recommended for monitoring every hectare for each pest species. More traps may be deployed for mass-trapping. The traps may be set up when the crop is 30 to 40 days old. A lure has to be replaced in the field once in 15 days or as recommended by the suppliers. However, in sealed packets the lures can be stored in the refrigerator for about 12 months. The lure, being species specific, is different for each pest species. When the traps fitted with lures are deployed in the field, the moths of the cotton bollworms or concerned pests get attracted to their respective lures and fall into the traps.

Yellow sticky traps: Yellow sticky traps can be used to monitor aphids and whitefly. In cabbage and cauliflower, light traps are useful in reducing diamond back moth incidence. A bulb (100 watts) can be fixed by hanging in front of a thin gunny bag or polyethylene sheet (3'x4') smeared with oil or grease. A bucket with water can also be kept below the bulb. About 5-8 light traps are to be set up per hectare.

Nylon net: Cultivation of vegetables and other crops under nylon net drastically reduces the pest populations.

Use of light trap: Light trap can be used to monitor and trap adult pests thereby reducing their population. Some common forms of light traps used are bonefires, electric bulbs and hurricane. A large plate or vessel containing kerosene mixed with water is placed near the light. The light trap should be 2-3 feet above the crop canopy. The trap should be set up in the field between 6-9 pm. After 9 pm, there are chances of beneficial insects getting trapped. Adult moths, which are attracted by the bright light, fall into this water and die.

Bird perch: The bird perch method involves the use of certain structures that invite birds to the fields when the larval population is high. 'T'-shaped bird perches can be erected in the fields at the rate of 3-5 per hectare. Turmeric powder mixed with rice can also be placed on the perches to attract birds. They should be one feet above the crop canopy. These perches serve as resting places for the birds that feed upon the larvae in the fields. Several species of insectivorous birds have been found to feed on insect pests of crops which are known to reduce the larval population to the extent of 70-80 per cent. Among the predatory birds, black drongo, house sparrows, blue jays, cattle egret, rosy pastor and mynah have been commonly recorded predators on large number of *H. armigera* and lepidopteran insects on chickpea, pigeonpea, and groundnut crops.

5.1.5 Other methods

Apart from plant extraction, there are some other natural pesticides that are allowed in organic farming. Although some of these products have limited selectivity and are not fully biodegradable, there are situations when their use is justified. However, in most cases, the desired effect is best reached in combination with preventive crop protection methods.

- Ashes: against soil-borne disease, ants, leaf miners, stem borers
- Slaked lime: against soil-borne diseases
- Clay: against fungal diseases
- Baking soda: against fungal diseases
- Soft soap solutions: against aphids and other sucking insects
- Light mineral oil: against various insect pests (harms natural enemies)
- Sulphuric acidic argillaceous earth: against fungal disease

Preparation of Bordeaux mixture

Dissolve 1 Kg of copper sulphate in 10 litres of water. In another vessel, slake 1 kg of quicklime by adding small quantity of water preferably warm water (1 to 1.25 kg of lime can be taken if the lime is not of good quality). When slaking is over, add 5 litres of water and stir well to get a uniform suspension of lime. Transfer the lime suspension through a sieve to a vessel containing 85 litres of water and stir well. A small quantity of lime solution may be kept separately.

Add 10 litres of the copper sulphate solution to the 90 litres of lime solution with constant stirring. To test the correctness of the mixture, dip a brightened iron knife for a minute in the mixture. If the knife remains bright, the mixture is correctly prepared. If the knife turns rusty brown or if its brightness is lost, add more lime suspension. Correctly prepared Bordeaux mixture will turn red litmus to blue and turmeric powder to orange red in colour.

Important points:

1. For dissolving copper sulphate or Bordeaux mixture, use copper, wooden or earthenware or plastic pots or drums.
2. Use fresh quicklime.
3. Bordeaux mixture should be passed through a sieve before transferring to the sprayers.
4. Spraying of Bordeaux mixture should be done on the same day of preparation.

Contamination Control: It is necessary to take some measures to minimise the contamination from outside and within farm.

- If neighbouring fields are non-organic, a buffer zone must be maintained.
- If the farm is under conversion, equipments used for conventional areas should be well cleaned before using for organic areas.
- Products based on polythene, polypropylene and other polycarbonates are allowed to cover protected structure, insect netting, nursery, drying, etc. subject to the condition that these materials should be removed from the field after use and they should not burnt or put in the soil. Use of polychloride based products like PVC pipe is prohibited.
- Farm equipments from conventional farming should be free from residues and clean.

The disease and pest management methods listed above can be integrated in to the organic farming systems depending upon the crops and the pest species. It is possible to keep the pest populations below economic injury levels by conserving existing natural enemies and adoption of suitable biological, cultural and mechanical methods.

5.2 Weed Management

Weeds can be considered a significant problem because they tend to decrease crop yields by increasing competition for water, sunlight and nutrients while serving as host plants for pests and diseases. It is important to understand that under an organic system of weed control, weeds will never be eliminated but only managed.

5.2.1 Cultural method of weed control

Crop rotation: Crop rotation involves alternating different crops in a systematic sequence on the same land. It is an important strategy for developing a sound long term weed control program. Weeds tend to thrive with crops of similar growth requirements as their own and cultural practices designed to contribute to the crop may also benefit the growth and development of weeds. Monoculture, that is growing the same crop in the same field year after year, results in a build-up of weed species that are adapted to the growing conditions of the crop. When diverse crops are used in a rotation, weed germination and growth cycles are disrupted by variations in cultural practices associated with each crop (tillage, planting dates, crop competition, etc). It is best to alternate legumes with grasses, spring planted crops with fall planted crops, row crops with close planted crops and heavy feeders with light feeders.

Cover crops: Rapid development and dense ground covering by the crop will suppress weeds. The inclusion of cover crops such as rye, red, clover, buckwheat and oilseed radish or over wintering crops like winter wheat or forages in the cropping system can suppress weed growth. Highly competitive crops may be grown as short duration 'smother' crops within the rotation. Additionally, cover crop residues on the soil surface will suppress weeds by shading and cooling the soil. When choosing a cover crop, consideration should always be given to how the cover crop will affect the succeeding crop. In addition, decomposing cover crop residues may release allelo chemicals that inhibit the germination and development of weed seeds.

Intercropping: Intercropping involves growing a smother crop between rows of the main crop. Intercrops are able to suppress weeds. However, the use of intercropping as a strategy for seed control should be approached carefully. The intercrops can greatly reduce the yields of the main crop if competition for water or nutrients occurs.

Field scouting: It involves the systematic collection of weed and crop data from the field (weed distribution, growth stage, population, crop stage etc). The information is used, in the short term, to make immediate weed management decisions to reduce or avoid economic crop loss. In the long term, field scouting is important in evaluating the success or failure of weed management programs and for making sound decisions in the future.

Mulching: Mulching or covering the soil surface can prevent weed seed germination by blocking light transmission preventing seed germination. Allelopathic chemicals in the mulch also can physically suppress seedling emergence. There are many forms of mulches available.

Such materials as straw, bark, and composted material can provide effective weed control. Materials such as black polyethylene have been used for weed control in a range of crops in organic production systems. Plastic mulches have been developed that filter out photosynthetically active radiation, but let through infrared light to warm the soil. These infrared transmitting mulches have been shown to be effective at controlling weeds.

Tillage system: Tillage systems alter the soil seed bank dynamics and depth of burial of weed seeds. Weed seedling emergence is often more uniform shallow buried weed seeds and may result in better weed control. Weed seeds closer to the soil are more likely to be eaten or damaged by insects, animals, other predators and disease causing organisms.

Sanitation: The use of clean seed, mowing weeds around the edges of fields or after harvest to prevent weeds from going to seed, and thoroughly composting manure before application can greatly reduce the introduction of weed seeds and difficult weed species.

Nitrogen fertility: Selective placement of nitrogen in a band can favour the crop over the weed. Use of legume residues are opposed to chemical nitrogen fertilizer to supplement nitrogen needs of the crop can enhance weed suppression.

Pre-germination of weeds: Irrigation or rainfall helps to germinate weed seeds just before the cash crop is planted. The newly germinated weeds can be killed by light cultivation. Pre-germination occurs as close a possible to the date of planting to ensure that changes in weather conditions do not have an opportunity to change the spectrum of weeds (cool vs. warm season) in the field.

5.2.2 Mechanical method of weed control

Mechanical removal of weeds is both time consuming and labour-intensive but is the most effective method for managing weeds. The choice of implementation, timing, and frequency will depend on the structure and form of the crop and the type and number of weeds. Cultivation involves killing emerging weeds or burying freshly shed weed seeds below the depth from which they germinate.

Mechanical weeders include cultivating tools such as hoes, harrows, and brush weeders, cutting tools like mowers and stimmers, and dual-purpose implements like thistle-bars. The choice of implement and the timing and frequency of its use depends on the morphology of the crop and the weeds. Implements such as fixed harrows are more suitable for arable crops, whereas inter-row brush weeders are considered to be more effective for horticultural use. The brush weeder is mainly used for vegetables such as carrots, beetroot, onions, garlic, etc. Hand hoes, push hoes and hand-weeding are still used when rouging of an individual plant or patch of weed is the most effective way of preventing the weed from spreading. Hand-weeding may also be used after mechanical inter-row weeding to deal with weeds left in the crop row.

Soil solarization: During summer, organic farmers sterilize their soil through solarization. In this process, a clear plastic film is placed over an area after it has been tilled and tighly sealed at the edges. Solarization works when the heat crated under the plastic film becomes intense enough to kill weed seeds.

Questions

1. Describe important preventive crop protection measures used in organic farming.
2. What are the three types of Bio-control procedures?
3. What are predators and parasitoids?

4. What do you understand by term "Antagonistic Organisms".
5. Describe role of biopesticides in controlling the pests of crops in organic farming.
6. What is pyrethrum? How it is used in pest controls?
7. What is sex pheromone? How pheromones helpful in the detection of pests and their timely control?
8. Describe cultural method of insect-pest control in crops.
9. What are the mechanical methods used in controlling the pests in crops? Give three examples.
10. Describe different methods of weed management in organic farming.
11. What is Bordeaux mixture? How Bordeaux mixture is prepared?
12. Fill in the blanks:
    a. Biological Control is defined as the reduction of pest populations by ....
    b. Chemicals used to control pests are called ....
    c. Disease causing organisms are called ....
    d. In plants, diseases are mainly caused by ....and ....
    e. Chemical pesticides are .... and .... in nature.
    f. .... is considered as the fastest growing and an alternative farming sector for safe food production process.
    g. The conversion of a minor pest to a major one, in the absence of natural enemies is termed as ....
    h. The ladybird beetles are mass-produced and made commercially available to the farmers in the form of .... or ....
    i. The pesticidal compounds derived from living organisms are called ....
    j. Yellow sticky traps can be used to monitor.................... and ....
13. Differentiate between followings :
    a. Trap crops and Light trap
    b. Sex pheromone and Bird perch

c. Cover crops and Intercrop

d. Parasitoids and Predators

14. Define following terms:

a. Curative crop protection

b. Biological crop protection

c. Cultural pest control

d. Mechanical pest control

CHAPTER - 6

Quality Control Standards and Certification

The organic quality is based on standards, inspection, certification and accreditation. All organic food is produced and handled according to strict rules called 'Organic Standards'. Organic standards do not define a quality status which can be measured in the final products (e.g. quantity of pesticides residues, heavy metals, etc). They define the way of production (e.g. that no chemical pesticides and fertiliser are used).

6.1 Important Organic Standards

IFOAM basic standards: The most important organic standards are the IFOAM Basic Standards which also describes the principle of organic farming and provide recommendations on how to achieve the minimum requirements. Being the 'mother of organic standards', IFOAM Basic standards are not standards for certification but "*standards for standard*" setting on the national or international level. It provides a framework for certification bodies and standard setting organizations worldwide to develop their own certification standards. They are regularly reviewed and updated in a democratic process by the IFOAM members from all over the world.

Codex Alimentarius guidelines: The *Guidelines* include general sections describing the organic production concept and the scope of the text; description and definitions; labelling and claims (including products in transition/ conversion); rules of production and preparation, including criteria for the substances allowed in organic production; inspection and certification systems; and import control. The Codex Alimentarius Commission at its 23rd Session in 1999 adopted the *Guidelines for the Production, Processing, Labelling and Marketing of Organically Produced Foods*, with the exception of the provisions for livestock and livestock products. The Codex Alimentarius Commission at its 24th Session in 2001 adopted the sections concerning livestock and livestock products and beekeeping and bee products for inclusion in the *Guidelines for the Production, Processing, Labelling and Marketing of Organically Produced Foods.*

EU Regulation (EEC) No 2092/91: The European Union (EU) was one of the first to set up a policy on organic farming by adopting EU Council Regulation (EEC) No 2092/91, amended by Council Regulation (EC) No 1804/1999. With this regulation, the Council created a community framework defining in detail the requirements for agricultural products or foodstuffs bearing a reference to organic production methods. The regulation is set up primarily as a labelling regulation, meant to regulate the internal market for organic products but it also describes the organic production standards and the inspection and supervision requirements. As it deals with virtually all agricultural products and with all aspects of primary food production and food processing, the remit of the regulation is very broad. At its creation in 1991, the regulation took into account, to a large extent, the existing private production rules. At that time, only a few Member States had developed national legislation. While the original regulation was relatively short and covered only plant production. In 1999, it was extended substantially to cover animal production in a relatively detailed way. Being a Council regulation, this legislation is directly applicable in all Member States.

Indian National Standards for Organic Production (NSOP): In India, standards for organic agriculture were announced in May 2001, and the National Programme on Organic Production (NPOP) is administered by Agricultural and Processed Food Products Export Development Authority (APEDA) using the IFOAM Basic Standards (Edition 2000) under the Ministry of Commerce. Definite principles, basic standards of production, documentation, inspection and certification guidelines are approved by the National Standards Committee constituted by the members of IFOAM in India. Certain responsibilities to define details are delegated to the accredited certification bodies.

The government has set the frame conditions in which the organic sector of a country operates. Most important are the content and legal status of organic standards, the regulations concerning the use of organic claims and labels, the legislation on consumer protection and the accreditation system. As per the national accreditation policy, all the certifying agencies operating in India are to obtain accreditation from anyone of the accrediting agency appointed by the Government of India, viz., Spices Board, Coffee Board, Tea Board or APEDA.

The Indian National Standards for Organic Production and India Organic Logo are governed by APEDA, which provides national standards for organic products through a National Accreditation Policy and Programme.

The aims of the National Programme for organic production include:

1) to provide the means of evaluation of certification programmes for organic agriculture and products as per internationally approved criteria.
2) to accredit certification programmes.
3) to facilitate certification of organic products in conformity to the National Standards for Organic Products.
4) encourage the development of organic farming and organic processing.

Other efforts towards promotion of organic exports include attempts to collaborate with all the major organic importing countries. The Certification according to NPOP is now internationally recognised. This is an added advantage for the export segment of Indian farmers. The latest copy of Indian organic standards can be procured from Agricultural and Processed Food products Export Development Authority (APEDA), Ministry of Commerce, Govt. of India, New Delhi.

6.1.1 Common issues in the various organic standards

Application of organic standards: The standard's requirements should be met during the conversion period. All the standard's requirements should be applied on the relevant aspects from the beginning of the conversion period onward. If the whole farm is not converted, the certification programme should ensure that the organic and conventional parts of the farm are separate and inspectable. Converted land and animals should not get switched back and forth between organic and conventional management.

Soil health and water quality: Necessary measures have to be taken to maintain and improve landscape and enhance biodiversity quality. Land preparation by burning vegetation is restricted to the minimum. Crop production, processing and handling systems should return nutrients, organic matter and other resources removed from the soil through harvesting by the recycling, regeneration and addition of organic materials and nutrients. Grazing management should not degrade land or pollute water resources. Relevant measures should be taken to prevent or remedy soil and water salinization. Operators should not deplete nor excessively exploit water resources, and should seek to preserve water quality. They should where possible recycle rainwater and monitor water extraction. Clearing of primary forest is not permitted. The farmer should take defined and appropriate measures to prevent soil and water erosion.

Biodiversity: Diversity in plant production and activity should be assured by minimum crop rotation requirements and/or variety of plantings. Minimum rotation practices for annual crops should be established unless the operator demonstrates diversity in plant production by other means. Operators are required to manage pressure from insects, weeds, diseases and other pests, while maintaining or increasing soil organic matter, fertility and microbial activity. For perennial crops, the certifying body should set minimum standards for orchard/ plantation floor cover and/or diversity or refuge plantings in the orchard.

Soil productivity: The productivity of the farmland should be preserved and promoted only by applying the compost derived from the remainders of the agricultural products produced in the said fields, etc. and methods effectively utilising biological functions of the organism inhabiting and growing in the fields. Material of microbial, plant or animal origin should form the basis of the fertility program. Nutrients and fertility products should be applied in a way that protects soil, water, and biodiversity. The fertility and the biological activity of the soil must be increased, in the first instance, by cultivation of legumes, green manures or deep rooting plants in appropriate rotation programme.

Mineral fertiliser (ground rock, lime, dolomite, etc), microbial fertilisers, biodynamic preparations and botanical preparations only to be used as supplement. No synthetic/chemical fertilisers are permitted. In cases where the productivity of the farmland cannot be preserved and promoted only by the methods utilising the biological functions of the organism inhabiting and growing in the said fields or in the circumference, utilise only the fertilisers and the soil improvement materials noted in the organic standards. The maximum amount of manures may not exceed 170 kg nitrogen/year/hectare of agricultural area used (EEC2092/ 91).

Pest, disease and weed management: Control of noxious animal, insects, pests, disease, weeds and plant in fields, etc. should be executed only by the cultivation method (suitable crop lists and variety, the adjustment of the cropping time, and other cultivation management of the agricultural products), physical method (using light, heat, sound, etc., or manual or mechanical methods), biological method (by introducing microorganisms suppressing the proliferation of microorganisms being the cause of diseases, predators, plants repelling noxious animal and plant, or by improving the environment suited for growing them), or an appropriate combination of these methods. In cases of being critical, the agricultural chemicals noted in the organic standards may be used. Preventive methods should be adopted to maintain the plant health. Ionising radiation should not be executed for the disease and pest control, the preservation of the foods, removal of pathogens or sanitation. The products which can be used under the supervision and inspection by certification body for pest and disease controls are for example: bio-pesticides (Bt, *trichoderma*, NPV, pseudomonas etc.), Bordeaux mixture, sulphur, soft soap, and most plant based products (neem, rotenone, pyrethrum, etc.). Maximum use of copper under NPOP is 8.0 kg/ha/yr. EU Regulation restricted the use of copper up to 6 kg/ha/yr on a 5-year average.

Contamination control: A buffer zones or border crops must be established to avoid potential contamination and limit contaminants in organic products. In the transportation, selection, processing, cleaning, storage, packaging, and other processes, control in such a manner as not being mixed with other agricultural products than the organic agricultural products. For synthetic structure coverings, mulches, fleeces, insect netting and silage wrapping, only products based on polyethylene and polypropylene or other polycarbonates are permitted. These should be removed from the soil after use and should not be burned on the farmland. All equipment from conventional farming systems should be thoroughly cleaned of potentially contaminating materials before being used on organically managed areas.

Collection of wild harvested product: Wild harvested products should only be certified organic if derived from a stable and sustainable growing environment. Harvesting or gathering the product should not exceed the sustainable yield of the ecosystem, or threaten the existence of plant or animal species. Products can only be certified organic if derived from a clearly defined collecting area, which is not exposed to prohibited substances, and which is subject to inspection. The collection area should be at an appropriate distance from conventional farming, pollution and contamination. The operator managing the harvesting or gathering of the products should be clearly identified and be familiar with the collecting area in question. The collection does not affect the stability of the natural habitat or the maintenance of the species in the collection area. The harvesting of wild crop should be in a manner that such harvesting will not destructing to the environment and will sustain the growth and production of the wild crop.

Animal management: The establishment of organic animal husbandry requires an interim period, the conversion period. Management of the animal environment takes into account the behavioural needs of the animals and provides for sufficient free movement, sufficient fresh air and natural daylight; protection against excessive sunlight, temperatures, rain and wind; enough lying and/or resting area; natural materials for all animals requiring bedding; ample access to fresh water and organic feed; adequate facilities for expressing behaviour in accordance with the biological and ethological needs of the species.

Uses of preventive antibiotics or growth promoters are not permitted. No chemicals (including urea) or artificial vitamins should be added to the feed. At least 50% of the feed should come from the own farm itself or produced in cooperation with other organic farms in the region. If certified organic feed is not available, a maximum of 20% of non-organic feed can be fed (max. 15% for ruminants), provided it does not contain prohibited chemical additives or genetically modified products. The percentage should be reduced within 5 years to 15% (10% for ruminants).

The use of conventional veterinary medicines (e.g. antibiotics) is allowed when alternative treatment is not sufficient. Synthetic growth promoters and hormones are not allowed. Embryo transfer techniques, hormonal heat treatment and genetically engineered animals are not allowed. Breeding systems should be based on breeds that can reproduce successfully under natural conditions without human involvement. Artificial insemination is allowed. Landless animal husbandry systems are not allowed. Herd animals should not be kept individually. The certification programme may allow exceptions e.g. male animals, smallholdings, sick animals and those about to give birth. Poultry and rabbits should not be kept in cages.

Breeding stock brought-in from conventional farms with a yearly maximum of 10% of adult animals of the same species on the farm.

Social justice: Social right and justice are integral part of organic agriculture. The laws relating to labour welfare and rights of children should be honoured. All employees and their families should have access to potable water, food, housing, education, transportation and health services. All employees should have equal wages when doing same job. They must have equal opportunities irrespective of colour, creed and gender. Social security needs (include maternity, sickness and retirement benefits) should be met. Labour conditions regarding noise, dust, light and exposure to chemicals should be within acceptable limits; and they should have adequate protection. The rights of indigenous people should be respected.

6.1.2 Some key issues in the various organic standards

In organic farming system, certain minimum requirements are to be met to fulfil its objectives. The requirements of all the Regulations are based on the same principles, there is slight variation from each other.

Conversion: When a farmer switches over to the system of organic farming from the conventional system of farming, it is known as conversion. The time between the start of organic management and certification is called conversion period. The conversion period is decided based on the past use of the land and ecological situation. In NPOP, Regulation (EEC) 2092/91 and most other standards, the conversion period for annual crops is 24 months and for perennial crops is 36 months from the start of organic management. However, the inspection activities must take place from the very beginning. The produce of the first 12 months has to be marketed only *as conventional*. The products from 12 to 24 months for annual and 24 to 36 months for perennials can be marketed as *'organic-in-conversion'* if a buyer wants such a product. Only after 24 months (for annuals) and after 36 months (for perennials) the products can be sold as organic with the relevant certificates. However, for USDA (United States) NOP, the conversion period is three years irrespective of whether the crop is annual or perennial and there is no in conversion certificate during the transition period.

Animal products to be sold as organic only after the farm or relevant part of it has been under conversion for at least 12 months. For dairy and egg production the animal production standard should have been met for not less than 30 days.

Planting material: Species and varieties cultivated should be adapted to soil and climatic condition and resistant to pests and diseases. According to all the regulations, planting materials must be from certified organic sources. Use of conventional seedlings/seeds is not allowed in Regulation (EEC) 2092/91 from January 2005. In case the organic planting materials are not available, conventional untreated seeds/planting materials can be used one time with prior approval from the certification body proving the unavailability of the product in the market (only accepted by NSOP, India). Use of genetically engineered seeds or planting materials such as tissue culture, pollen culture and transgenic plants are not allowed.

Mixed farming and cropping pattern: Animal husbandry, poultry, fisheries, etc. should be practised in addition to agricultural farming. Shifting cultivation is not allowed. Crop rotation should be followed if annual crops are grown. Intercropping should be practised when perennial crops are grown. Crop rotation should cover green manure as well as fodder crops. In case of perennial crops, cover crops should be grown to protect the soil. Mono-cropping should be avoided. Per acre of farm land, at least 3 trees must be grown. Farmers should avoid overgrazing and over manuring of land by keeping appropriate number of animals per farm land. Maximum number of animals per acre of farm land is: 2 milk cow or buffalo/ha, or 5 calves, or 13 sheep or goat, or 14 pigs, or 580 chickens.

Soil and water conservation: Measures like stone pitching/contour wall construction are to be taken up to prevent soil erosion. In case of saline soils, saline resistant varieties may be grown. Judicious irrigation is to be practised. Mulching is required. Pollution of surface and ground water should be prevented. Clearing of primary forest is prohibited. Cleaning of land through straw burning should be restricted to minimum.

Parallel production: Parallel production is growing the same variety or indistinguishable variety of crops by the same operator in different qualities. This is strictly not permitted for annual crops. In case of perennial crops the entire unit will have to be converted in five years. During this period, strict separation must be maintained and clearly documented. This is not a major concern in case of USDA-NOP as long as sufficient separation is maintained.

Documentation: Documentation of farm activities is must for acquiring certification when organic crops are raised. The following documents/records are to be maintained.

a) Field map	e) Output record	i) Pest control records
b) Field history sheet	f) Harvest record	j) Movement record
c) Activity register	g) Storage record	k) Equipments cleaning records
d) Input record	h) Sales record	l) Labelling records.

Processing, grading, labelling and packaging: Processing technologies like solar drying, freeze-drying, hot air chambers are permitted. Irradiation of agricultural produce is not permitted. No synthetic additives are to be added during processing.

Japan (JAS) standard requires grading procedure. Grading is an internal verification that a specific lot of product meets the JAS standards and that internal operational procedures were followed in production before the JAS seal is affixed.

The label should convey clear accurate information on the organic status of the product. (i.e. conversion in progress or organic). The labels for organic and conversion in progress products should be distinguishable by different coloured labels. The details like name of the product, quantity of the product, name and address of the producer, name of certification agency, certification, lot number etc. are to be given in the label.

For Example,

Information required on the label	
Crop	OG (Organic Ginger)
Country	I (India)
Field No.	05
Date of harvest	32 (1st Feb)
Year	2009
Lot number	OGI 0532 2009.

Lot number is helpful in tracing back the product particularly the field number in which it is grown in case of contamination. Lot number should include the crop, country, field number, date of harvest (in Julian calendar) and production year.

For packing, recycling and reusable materials like clean jute bags, should be used. Use of biodegradable materials should also be used. Unnecessary packaging material should be avoided. Organic and non-organic products should not be stored and transported together except when labelled.

Export: Each individual consignment/invoice requires a transaction certificate. This achieves two objectives. Firstly, the certification agency is able to keep track of the material produced and marketed from each certified production unit. Secondly, the organic quality of the product is proved at the point of entry into the country of import. In EU, import permission has to be obtained for each product for every specific producer, exporter and importer.

6.2 Certification Process

The difference between organic farming and other sustainable system like eco-farming, permaculture etc. is that the former is characterised by certification, which is a procedure bound by rules and regulations. All products sold as "organic" must be certified. The farmer, processor/manufacturer, distributor and retailer have to go through a process of certification. Certification for organic farms is an essential pre-requisite for marketing their produce, especially in the international markets. This is done mainly to ensure genuineness of organically farm products reaching the market and consumer. The certificate issued by an accredited agency is the attestation of the organic method of production of the products.

Any person or organisation intending to produce or process organic products must be subject to an inspection and certification procedure by an approved inspection body. Certified organic refers to agricultural products that have been grown and processed according to uniform standards, verified by independent state or private organizations accredited by the standard setting organisation. Anyone contravening this Regulation could be subject to prosecution by the Trading Standards Officers. The labelling and marketing of organic products are controlled by the National Regulation.

The certification process actually starts in the conversion process. The steps to be followed for certified organic production are as follows:

- Fulfil the requirement of conversion period.
- Develop a comprehensive operations plan for the farm including an Ecosystem Management Plan.

- Convert the farm completely or comply with strict regulations in the cases of split and parallel operations (organic and conventional).
- Farm follow-up inspections by certifier.
- Continuously update the chosen certifier on major changes.
- Keep track of inspection certificates issued by certifier.

Procedure of certification: The producer or farmers makes contact with certifying agency. Certification agency provides information on standards, fees, application, inspection, certification and appeal procedures. The producer then submits application along with field history, farm map, record keeping system etc. The contract indicating scope, obligation, inspection and certification, sanction and appeals, duration, fee structure is executed. The costs of certification depends on size of farm, type of production system, group of farmers, extent of animal husbandry, location of unit, travel time to reach the inspection site, and costs for travel, food and accommodation during inspection.

Inspection: Inspection plays significant role in certification mechanism. If an organic farmer wants his produce to be certified, he/she has to follow the norms and allow an inspection by a certifying agency. The inspecting agency evaluates the performance of the farm activities with the help of farmers' statements and records; and by viewing the fields, animals, practices followed, condition of farm building etc. He checks whether the statements and records are correct and plausible. On the basis of inspection report, the certification agency decides whether certification may be granted or denied.

Organic farmer has to undergo an inspection at least once a year. Inspectors verify that organic practices such as long-term soil management, buffering between organic farms and neighbouring conventional farms, and record keeping are being followed. Processing inspections include review of the facility's cleaning and pest control methods, ingredient transportation and storage, and record keeping and audit control. The inspector evaluates the performance of the farm activities with the help of the farmer's statements and records and by viewing the fields, animals and farm buildings. He can take samples for laboratory testing and may conduct unannounced inspections. The inspector transmits his/her findings to the certification body as a written report.

Certification: A defined procedure in which a certification body assesses a farm or company and assures in writing that it meets the requirements of the organic standards. Certification includes annual submission of an organic system plan and inspection of farm fields and processing facilities. The certification body compares the results of the inspection with the requirements of the organic standards. A certification committee decides whether certification may be granted or not and then the agency issues approval or denial of certificate. **Certificate**

is given for current year's harvest only and hence annual certification is required. In case of denial of the certificate, if the operator has valid reasons not to accept the certification decision, he/she can request for reconsideration of the decision in writing.

Accreditation: In order to make sure that the certification programme is competent to carry out inspection and certification, a third level of quality control is needed. Authorised bodies regularly evaluate certification programmes and check their proper functioning according to certain criteria. In case the certification body complies with the criteria, they accredit the certification programme.

In India, there are, at present, six accreditation agencies approved by the central government's Ministry of Commerce (MoC). They are the Agricultural and Processed Food Products Export Development Authority (APEDA), Coffee Board, Spices Board, Tea Board, Coconut Development Board and Cocoa & Cashew nut Board. Several other Indian initiatives have also applied for accreditation. Increasing competition can be expected in the Indian certification market in the near future.

6.2.1 Type of certification

Foreign certification: Presently, a lot of the export oriented organic projects in developing countries are inspected and certified by certification bodies based in the importing countries. These international certification bodies usually have long experience in organic agriculture and its certification. Importers often prefer their services as they are well known and provide their services worldwide. The main disadvantage is that costs of certification are high due to frequent plane trips and western salaries have to be paid.

Co-certification: During the last few years, most Western certification programmes started to build up local branch officers for conducting the inspections and to work with local inspection staffs. Still, the inspection work is supervised by the head office, but the number of required plane trips is less. Local Inspectors find it easier to inspect farms, as they usually speak the same language and are familiar with the local conditions.

Indigenous Certification: Indigenous certification bodies can usually offer cheaper inspection fees as less travelling is required and only local salaries have to be covered. Indigenous certification may especially support the development of domestic market for organic products. However, for export purposes local certification bodies have to achieve international recognition which means to meet the different requirements of different import countries.

In 2002, a group of farmer organisations, NGOs and companies established an Indian Organic Certification Agency named INDOCERT as a charitable trust in order to provide affordable certification services of high quality. Under the guidance of FiBL and bio.inspecta AG and with support from SECO (State Secretariat for Economic Affairs), the office started offering inspection and certification services in August 2002. Since then, even the rates of the international certification bodies have come down significantly. Through re-certification arrangements with bio.inspecta, INDOCERT certification from the beginning enable access also to export markets. Recently, INDOCERT is accredited by the DAP, Germany for certification for EU Regulation 2092/91. This certification agency is the first fully Indian Certification Body listed by FOODPLUS to offer EUROGAP Certification in India. There are some other indigenous certification initiatives, but so far none of them gained momentum.

It must be understood that, for export to the EU market, a certificate from a certification body, which is accepted by EU competent authorities, is essential. To regulate the export of certified organic products, the Director General of Foreign Trade, Government of India has issued a public notice according to which no certified organic products may be exported unless they are certified by an inspection and certifying agency duly accredited by one of the accreditation agencies designated by the Government of India. For the domestic market also it is expected that certification will be made compulsory.

Certification Bodies in India

- ECOCERT International (based in France and Germany, branch office in Aurangabad, Maharashtra).
- IMO Control Pvt. Ltd.–Institute for Marketecology (based in Switzerland, office in Bangalore, Karnataka).
- LACON GmbH (based in Germany, office in Aluva, Kerala)
- SGS India Pvt. Ltd. (based in Switzerland, offices in Delhi and other cities).
- BIOINSPECTA (based in Switzerland, branch office in Cochin, Kerela)
- SGS India Pvt. Ltd (based in India, office in Banglore)
- APOF Organic Certification Agency (AOCA) (based in India, office in Gurgaon, Haryana)
- SKAL International (based in the Netherlands, branch office in Mumbai).
- INDOCERT (based in India, office in Aluva, Kerala)
- India Society for Certification (ISCOP) (based in India, office in Coimbtore).

Smallholder Group Certification: Where large numbers of smallholders are to be inspected by a foreign certification body, the involved costs can be very high. In such cases, smallholder group certification can be done. It is done for defined groups (may be up to 900-1000 farmers) of collaborating organic producers, processors, and exporters with similar farming and production system located in geographical proximity. Only one application is required to certify the entire farms of such smallholder group and the certification and inspection fee is shared by each individual farmer/operator. It should be based on internal regulations and defined sanctions in case of non-compliance. The group certification necessarily insists on internal control system (ICS).

The **Internal Controls System** operates like a small internal control body: internal standards, a written commitment of the participating farmers, internal inspectors inspecting the farms at least once a year and an internal system of sanctions against defaulting farmers. This is a documented quality assurance system that allows the certification body to delegate the inspection of individual group members to a body identified from within the operator group. The external certification programme evaluates the proper functioning of the ICS and re-inspects at random a certain percentage of the farms. Contracted party is the farmer group, project or corporate which also is the owner of the certificate. The external auditor can invariably insist on physical inspection of all individual holdings extending up to and above four hectares. It may also be noted that the number of units, above four hectares should not exceed 50 per cent of the total area under the group certification.

The following criteria may be used for identification of farmers for group certification:

Marginal Farmers	: Below 1 hectare
Small Farmers	: 1.00 – 1.99 hectare
Semi medium Farmers	: 2.00 – 3.99 hectare
Medium Farmers	: 4.00 – 9.99 hectare
Large Farmers	: Above 10.00 hectare

6.2.2 Hazard analysis critical control points (HACCP)

The Hazard Analysis Critical Control Points (HACCP) for food safety management has become an essential requirement of international trade to ensure higher confidence in food safety.

HACCP is a systematic process that identifies food safety hazards, critical control points (CCP), critical limits, corrective actions and documentation and integrates monitoring procedures to assure food safety.

Organic Critical Control Point (OCCP): It is a procedure in the system where loss of control may result in a loss of organic integrity. This approach is useful for organic auditor to identify and assess contamination and assess risks to organic products. It can also be successfully used by organic producers and processors to protect organic integrity and develop their organic plan.

The concept of OCP is to check the critical areas of operators for compliance. The steps to implementing an Organic Control Point System are:

a) Identify Organic Control Points, using farm maps, facility diagrams and product/process flow charts.

b) Establish Critical Limits for each OCP in the line of the standard.

c) Establish "Protective actions" for each OCP to prevent contamination or commingling.

d) Establish corrective actions for all products that have lost their organic integrity.

e) Implement an effective record keeping system to document OCP activities.

f) Monitor the effectiveness of the system through annual (or more frequent) inspection and evaluation.

OCCP at different stages: Once the Organic Critical Control Points have been identified, it is necessary to establish and document for each OCCP. The OCCPs at different stages are as listed below:

a) *OCCPs in Crop Production Parallel production (Organic and Conventional production of a crop/variety).*

- Split production
- Reliance on off-farm inputs for fertility management.
- Use of treated seeds (or GMO seeds) or seedlings.
- Use of irrigation water (origin).
- Measures taken to improve soil fertility or to prevent soil erosion.
- Impact of prohibited products on the neighbouring fields (Drift? Mixing of products?).
- Sharing equipment with non-organic farmers.
- Off-farm Storage area.
- Inadequate or Contradictory records.
- Selling more than was produced.

b) *OCCPs in Animal Husbandry*

- Appropriate stocking rates. (Over grazing? Water pollution?)
- Sufficient Access to pasture or free run.
- Off-farm sources or livestock.

- Purchase of feed (Conventional/Organic? Additives?)
- Use of antibiotics (Preventive use? Growth promoting?)
- Use of insecticides and/or parasitized.
- Drinking water contamination.
- Physical mutilations of animals.

c) *OCCPs in processing and trade*

- Organic origin of goods and ingredients.
- Production/trade of organic and conventional products.
- Storage of organic and non-organic ingredients in the same area.
- Permitted production methods.
- Permitted processing aids.
- Sanitation chemicals and procedures? Pest Management.
- Permitted packing material? Appropriate waste management.
- Flow of products. Correct amounts'?
- Appropriate labeling.
- Social conditions.

6.2.3 Cost of inspection and certification

Most certifiers are charging inspection and certification fees based on the number of person-days involved, plus fees for the issue of certificates. Sometimes, different fees are applied for small farmers, large farmers, and processors or traders. All the accredited certifiers publishes tariff for inspection and certification. The rate varies between the types of growers.

Certification Procedure

- After mutual correspondence with the certifying agency, and once convinced the farmer can pay 50-75 per cent of the inspection and certification cost as advance along with the filled in application form. The tariff on inspection and certification varies between agencies and between the types of farmers.
- On receipt of the application, the certifying agency would review it and send an offer to the farmer/farmer's group. If the grower accepts the offer it need to be signed and send to the certifying agency.
- On acceptance of the offer, the certifying agency will send an 'Organic

system plan' form. These organic system plans accompany the application to enable the certifying agency to review it for understanding the ability of the party to comply with the requirements. *Incomplete application will delay inspection.* On receipt of the completed application along with 'organic system plan' the inspection visit will be planned.

- The farmer must keep all the details documented with respect to the organic cultivation. This includes crop wise input-output details, farm map, details on chain of custody, sanction-compliance reports etc. The inspector on inspection may if required can ask for some amendments of the activities by the farmer, for compliance under intimation to the certifying agency.
- The inspection reports with compliance report are submitted to the quality manager of the certifying agency. The quality manager scrutinizes the report and suggests further compliance if any required. The operator/farmer need to pay the balance amount for the inspection and certification cost.
- On obtaining formal compliance report and balance fee, the farmer's file will be given to a certification committee with expert members. They review every step and consider certification based merits. Once cleared, the certifier issues a certificate which may *contain requirements if any for further compliance.*
- The certificate is issued with validity for a period of one year.

Over the past few years, many international certifiers opened branch offices in India, operated by Indian staff. Thus, the costs for certification came down considerably. At the same time, local certification bodies started to emerge, partially with the aim of further reducing certification costs. It can be assumed that certification costs will soon reach the lowest possible level while still ensuring quality requirements for inspection and certification work.

Organically certified export items requires 'transaction certificate' for each transaction. This needs to be obtained by the exporter. Similarly processing/ packing unit also needs to be certified before it is operationalized.

The quality system is administered through established documented procedures which are periodically audited to ensure that production, processing, handling, management, certification, accreditation, and other systems to meet the specified requirements and outcomes by standardised protocols.

6.2.4 Views on national program on organic certification

At present, India is in the nascent stage of organic farming. Hence, the Government of India has planned a National Program for Organic Production and Certification. The objective of the program is to provide an institutional mechanism for implementing national standards for organic products through a National Accreditation Policy and Program. Some of the favourable responses of the national government are-

- Comprehensive policy, a recognised by international certification agencies.
- Attracts more certification agencies in India; this will reduce the cost for certification, which is a major constraint for small farmers.
- Different agencies have different parameters, therefore standardisation would help.
- Certification bodies should tie up with other certification agencies in other countries.
- National Certification would centralise the certification procedure.

Cost and quality emerge as the major constraints for certification. Implementation of national certification agencies that should provide comprehensive policy prices (reduction in cost of certifications), international standardisation and simplified certification procedure with less documentation are some of the important strategies of the national government to improve/ promote the organic farming in India.

Questions

1. What does "Certified Organic" mean?
2. Who regulates the organic standards?
3. What is IFOAM and why IFOAM basic standards are known as "Standards of Standards"?
4. Briefly mention important organic standards for organic production.
5. What is an Internal Control System (ICS)?
6. What do you understand by co-certification and why it is considered useful for a producer?
7. What is accreditation? Who accredits the certification bodies under NPOP for certifying organic products? Give name of three accrediting agency appointed by the Government of India.

8. Why inspection of organic production system is required? What is important farming aspects covered in the inspection process?

9. What are key issues in the various organic standards?

10. Name five certification bodies of other countries operating in India?

11. Define the following terms :

    a. NSOP

    b. OCCP

    c. HACCP

12. Expand the following.

    IFOAM, APEDA, NPOP, EU

13. Fill in the blanks

    a. Organic quality is based on ...., ..., ...and ....

    b. All organic food is produced and handled according to strict rules called ....

    c. The most important organic standards are the ....Basic Standards.

    d. All the certifying agencies operating in India are to obtain .... from anyone of the accrediting agency appointed by the Government of India.

    e. The standard's requirements should be met during the .... period.

    f. The .... operates like a small internal control body in smallholder group organic certification.

    g. The .... for food safety management has become an essential requirement of international trade to ensure higher confidence in food safety.

    h. If an organic farmer wants his produce to be certified, he/she has to follow the norms and allow an .... by a certifying agency.

    i. All products sold as "Organic" must be ....

    j. The European Union (EU) was one of the first to set up a .... on organic farming.

CHAPTER - 7

Organic Market Opportunities

The market of organic products is growing globally every year. This organic market expansion makes it possible for farmers to reap the benefits of a trade with relatively high price premiums. Organic farming is carried on in almost all countries of the world. The share of organic farms in overall farmland and in terms of numbers of holdings is growing continuously. The market for organic products is also growing very rapidly, not only in Europe, North America and Japan, where the largest markets are to be found, but also in many emerging economies in transition. In India, an organic movement is now emerging on different levels i.e. producer groups, trainers and advisors, certification bodies and processors and traders.

7.1 Important Organic Products

Organic food products encompass an enormous diversity of product groups. The main product desired by European, American and Asian countries are- tea, coffee and cocoa; grains, pulses and seeds; vegetable oils and fats; edible nuts; spices and herbs; dried fruits; fruit juices and concentrates; sugars and honey.

Tea: Tea is traded as black tea, green tea, Oolong tea and instant teas.

- Black tea: Fully fermented tea.
- Green tea: Certain enzymes in the fresh leaves are inactivated through heat treatment. Only then is the product rolled and dried. Fermentation is suppressed by deactivating these enzymes and the leaves retain their green colour.
- Oolong tea: The fermentation process is halted at an earlier stage (partly fermented tea).
- Instant tea: Instant teas are made either from low-quality teas (fermented and dried), or from non-dried tea in a special process directly after fermentation. Instant teas lose much of their aroma

during the extraction and subsequent freeze-drying processes.

Coffee: Economically the most important coffee varieties are *Coffea arabica* (Arabica) and *Coffea canephora* (Robusta). In comparison with Arabica, 30 percent higher yields are gained from Robusta, but the price is around 30 percent lower. Raw coffee is made by processing the ripe red coffee cherries of the bush-like coffee tree. Blending and roasting the raw coffee is mostly carried out in the importing countries. Processing includes dry or wet processing, dehulling, grading, cleaning, sorting and filling.

Cocoa: With its strong, fine flavour the Criollo group produces the highest cocoa quality. However, its yield is low and it is therefore rarely cultivated. The Trinitario is a hybrid of Forastero and Criollo types. Its share of world production is roughly 10 to 15 percent.

Grains: USA and Canada dominate the market for organic commodity grain products (e.g. wheat, maize and barley) and opportunities for exporters in developing countries to export these grains are limited.

Pulses: It includes kidney beans, chickpeas and broad and horse beans. Although the trade is small, they are important for exporters from developing countries.

Seeds: It includes polyunsaturated oils, sunflower seed, oil and meal.

Vegetable oils and fats: Soya oil, palm oil and coconut oil which are important for exporters in developing countries.

Edible nuts: There are two segments for edible nuts, groundnuts (peanuts) and luxury (tree) nuts. The most important for the European trade are almonds, coconuts, cashew nuts, walnuts and Brazil nuts.

Spices and herbs: The main international trade for spices and herbs is dried and in crude form, cleaned but not further processed. It is estimated that about 85 percent of the trade is in this form. The remainder is for crushed or ground spices, essential oils or oleoresins.

Fruit: Apples, apricots, bananas, dates, figs, mangoes, papayas, peaches, pears and prunes are important tree fruits. Next to vine and tree fruit, there are others such as pineapple. Dried fruits are also in demand because fresh fruit consisting of more than 80 percent of water and prone to microorganism attack. Drying of fruit mainly done in order to stop the multiplication of microorganisms. Dried fruit can be divided into tree fruit and other fruit.

Fruit juices and concentrates: Fruit juice is orange juice. Apple, pineapple and grapefruit are other fruit species which are the basis for popular fruit juices.

Sugars and Molasses: The sugar that results from sugarcane and sugar beet are important as it contains the same sucrose that is found naturally in the original plants and in fruits and vegetables. A by-product of the cane and beet sugar refining process is molasses, which has a multitude of uses. Molasses is important as a raw material for the production of antibiotics, bakers yeast, rum and alcohol, as well as an animal feed supplement.

Honey: Five common types of honey have export market. They are - Acacia (a honey with a light taste and refined scent and tends not to crystallize); Orange blossom (a honey with a refreshing bittersweet flavour); Buckwheat (a honey with a strong smell and a taste similar to brown sugar); Lotus (a honey with a mellow, sweet flavour and a faint smell of flowers); Clover (the most widely produced and well-known type of honey)

7.2 The Indian Organic Market

In India, organic consumers are generally found in the urban upper-middle class or upper class, and to some extent lower-middle class families in smaller towns, especially families with children. Organic vegetables and fruits are the major organic products desired by Indian customers. A list of organic products desired by organic food consumers of India is presented below:

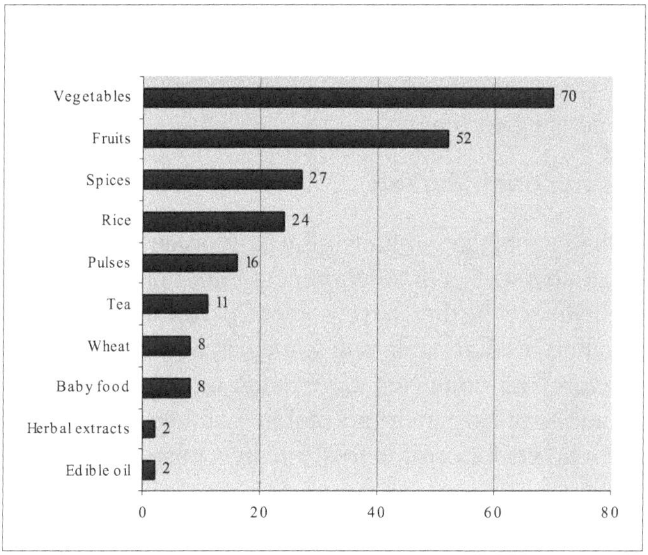

Organic Products desired by Organic Food Consumers

The major organic food products available in the Indian market are rice, tea, fruits and vegetables, wheat, spices, coffee and pulses. Major domestic markets are cities like Mumbai, Bangalore, Delhi, Chennai and Hyderabad to name a few.

Distribution of organic products takes place as follows:

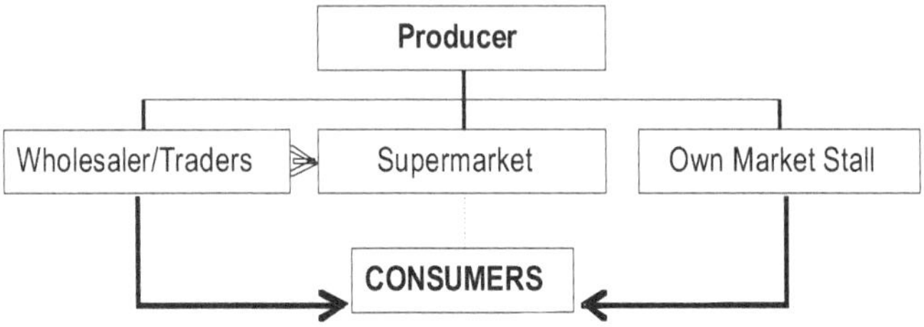

7.3 Export Market of Organic Products

Indian organic producers and exporters are well aware of the demand for organic products in developed countries. India's exports of organic food item handled by Agricultural and Processed Food Products Export Development Authority (APEDA) have increased from an estimated Rs.26 million US$ (1999) to an estimated level of Rs.100.4 million US$ (2003).

Products available for the export market are rice, wheat, tea, spices, coffee, pulses, fruits and vegetables, cashew nuts, cotton, oil seeds and medicinal herbs. The channels adopted for the export of organic products are mainly through export companies. The major export markets are Australia, Belgium, Canada, France, Germany, Italy, Japan, Netherlands, Sweden, Singapore, South Africa, Saudi Arabia, UAE, the United Kingdom and USA. Export sales of selected organic products are given below:

7.4 The International Market

One of the factors that promote growth in organic markets worldwide is consumer awareness of health, environment issues and food scandals. Other factors that influence further development of the organic market are the increasing promotions and marketing strategies used by key players, such as retailers. Developing countries are expanding their organic market into developed countries and in parallel are building a domestic market. An overview of world food markets for organic food and beverages (retail sales in 2003) is given below:

Country	Retail sales
Germany	2800-3100 million US$
Europe	10000-11000 million US$
US	11000-13000 million US$
Canada	850-1000 million US$
Japan	350-450 million US$
Worldwide	23000-25000 million US$

7.4.1 The Japanese organic market

The largest Asian market for organic products is found in Japan. The Japanese organic market has been characterized as a market with high demand for organic products and strong purchasing power, and with low domestic supply of organic products. The annual growth rate of the organic market is about 20 per cent.

The major exporters of organic products are Australia, New Zealand, the USA and Canada. The most commonly imported organic products are soybean, organic frozen vegetables, tea and bananas. According to the Japanese Integrated Market Institute, imports of organic products are likely to grow by 40 per cent.

7.4.2 The US organic market

The US is the world's largest market for organic products. Developing countries have sales opportunities in the following product categories:

- Tropical products (coffee, cocoa, tea, tropical fruits and vegetables)
- Off-season products (fruit and vegetables)
- In-season products (fruits and vegetables in temporary or permanent supply)
- Specialty and other products (wines, ethnic foods, herbs, spices, essential oils, sugar, feed grains, and seed grains)

To be certain that organic products will be accepted in the US organic market, the producer or exporter should choose a certification body that provides access to the US market, for example, a certification body accredited by the National Organic Program (NOP) of US; alternatively it is possible to have organic products re-certified by an accredited certification agency.

7.4.3 The European organic market

The European market for organic food is undergoing strong expansion. However, its development varies from country to country. Germany has the largest market volume within the EU, but the market in the United Kingdom is also growing strongly. Switzerland and Denmark have the highest consumption per capita, whereas France, Holland and Italy have low consumption per capita. Some domestic organic products already have a 15% share of the total market in Europe. In Germany and Switzerland one of the biggest challenges is to increase the diversity of the organic product range, and expand sales channels.

7.4.4 Demand of organic produce by International market

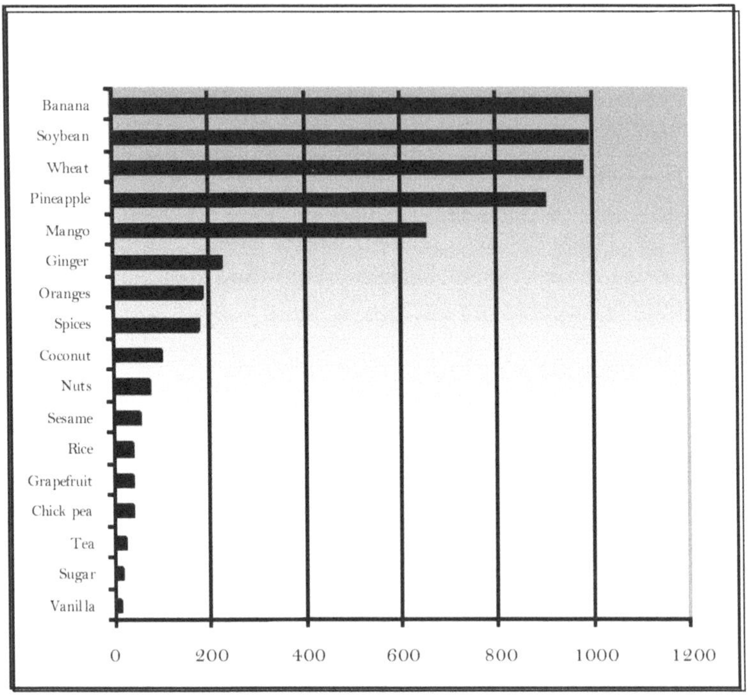

Demands of organic food in International Market (tons/year)

Of the products mentioned in the chart above, bananas are sold in high quantities followed by wheat and soybean. Tropical organic fruits such as pineapple and mango occupy third place. The rest of the organic products mentioned are sold in quantities ranging from 10 to 260 tonnes per year. The suppliers of these products are from neighbouring countries like Sri Lanka, China, Thailand, South America, Africa and Europe.

7.4.5 General price policies and cost structures

In general, an organic marketing initiative is expected to build in or carry more costs in contrast to conventional marketing. This is largely because of the 'additional' environment management and social responsibility it wants to carry. Organic marketing initiatives therefore have a distinct cost disadvantage against conventional businesses, which do not internalise equivalent environmental and social costs, in the market for similar products categories. An average price premiums in the export markets are given below:

Table 7.1 : Price premiums obtained in the export markets

Organic Product	Price premium %
Tea	47
Rice	44
Protein grains	44
Spices	30
Vanilla	53
Mango	25
Pineapple	28
Bananas	31
Nuts	40

Source : FiBL, 2002

Price Premium indicated by some exporters in the Asian Market is given below:

Exporter	Price Premium
Agri-Bio, Singapore	Retail price organic 3.3 times higher than conventional. Wholesale price is 40% higher than cost price and retail price is 85% higher.
Hanoi Organics, Vietnam	Retail price 40 - 120% higher than farm-gate price, depending on type of customer.
Keystone, India	Profit margin of 25-30%.
Nanjing Planck, China	Price organic vegetables 50 – 150% higher (100% on average) than conventional vegetables.
Organic Health, Malaysia	Profit margin is between 20 - 50%, but most items fall between 30 - 35%
Thai Organic Food, Thailand	Profit margin is about 30% to the cost of production, handling, and packing.

Australia, Argentina, Italy, the US and Brazil are respectively the five countries with the most cultivated area that is certified organic. Although, Asia has a very active organic movement (around a quarter of IFOAM's 703 members are from Asia), the area under organic cultivation remains relatively small. Among the more significant countries producing organic products are China, India, Indonesia, Sri Lanka as well as Israel. This is comparable to Canada and France, but smaller than Germany, Italy or the USA.

In most Asian countries, the local markets are just emerging. In almost all Asian countries, local distribution is a huge problem and often a price premium cannot be achieved. A number of countries, such as Israel, Japan, Malaysia and the Philippines have specialised organic stores or markets.

The Indian organic products, quantity and their domestic and export market are briefly indicated below in Table 7.2.

Table 7.2 : Indian organic products and markets

Product	Quantity (tonnes/yr)	Season	Markets
Tea	3500	Throughout the year	Domestic: Mumbai, Bangalore, Delhi, Hyderabad Export: Australia, Germany, Japan, Netherlands, UK, USA.
Coffee	600	Monsoon (June to September)	Domestic: Bangalore, Chennai, Hyderabad, Mumbai and Delhi Export: Australia, Germany, Japan, Netherlands, Sweden, UK, USA.
Spices	700	Throughout	Domestic: Bangalore, Chennai, Hyderabad, Mumbai, Delhi Export: France, Germany, Japan, Netherlands, South Africa, Singapore, UAE, USA
Rice	3500	Kharif: April to September Rabi: November to February	Domestic: Bangalore, Chennai, Hyderabad, Mumbai, Delhi Export: Japan, Singapore, UAE,USA, Canada, Germany.
Wheat	1400	Kharif: April to September Rabi: November to February	Domestic: Mumbai, Bangalore, Delhi, Hyderabad Export: Australia, Germany, Japan, Netherlands, UK, USA.
Pulses	400	March to May	Domestic: Mumbai, Chennai, Bangalore, Delhi, Hyderabad Export: Japan, Singapore, Germany, UAE, Saudi Arabia.
Oilseeds	100	Kharif April to September	Domestic: Negligible Export:European Countries
Fruits & Vegetables	2500	Throughout the year	Domestic: Mumbai, Chennai, Bangalore, Delhi, Hyderabad Export: UK, USA, Australia, France, Germany, Italy, Switzerland, Netherlands

Cashewnut	375	March to June	Domestic: Nil Export: European Countries
Others: Cotton, medicinal herbs & extracts, aloe vera, sapota	Cotton 1200 Herbs 250	Throughout the year	Domestic: Mumbai, Bangalore, Delhi, & Hyderabad Export: Australia, Belgium, Germany,Switzerland, Italy, Japan, Netherlands, UK, USA

Source : Org-Marg, 2002 (Source: Fieldwork, various publications from-APEDA, Tea Board, Spice Board, Coffee Board).

7.5 Global Organic Trade

Trade with organic products are showing growth rates over 20% which are rarely found in food markets. STRONG OPPORTUNITIES for developing countries in most major markets for organic products that are NOT PRODUCED in EUROPE or NORTH AMERICA, such as coffee, tea, cocoa, spices, tropical fruits, vegetables and citrus fruits. Organic products from developing countries are sold at impressive premiums, often at prices 20% higher than identical products produced on conventional farms. Although the current organic market is small, this is a big task. It is expected that once national standards are more widely established, bi-lateral equivalency agreements will lead to increased trade opportunities.

7.6 Limitations of Domestic Organic Market in India

A number of constraints have been identified in the marketing of organic produce. These are:

- Inadequate knowledge of the package of practices for organic production.
- Technologies are available only for a limited number of crops.
- Farmers do not know where to sell the organic produce and consumers do not know where to get genuine organic product.
- Market is limited and supply is inconsistent.
- Apathy of farmers to grow crops organically because of the low demand in domestic markets and non-realization of premium prices.
- Most farmers cannot distinguish organically grown items and the prices to be charged.

- Lack of consumer awareness on the benefit of organic produce.
- The quality supply is not ensured.
- The cost towards certification is fairly high. Farmer experiences difficulties in reaching certification agencies.
- Logistic of slow shipment for exporting Indian organic products.
- Time consuming and complicated paper work while dealing with export authorities.
- Inadequate efforts to develop domestic market.

7.7 Strategy to be taken for Market Improvement

Steps for creating domestic market :

- Development of crop-specific/location-specific organic package of practices that give details of the type of seed, variety, cultivation practices, nutrition, plant-protecting measures, harvesting stage etc.
- Development of rural market or 'haat' like Ryot Bazar (Andhra Pradesh) or Uzhavar Shanthai (Tamil Nadu) or 'Apni Mandi' where the arrangement for farmers-to-consumers direct selling can be made to ensure realization of a premium price.
- Development of home-delivery system.
- Registration of consumers wanting supply of organic foods.
- Marketing of organic produce in a vehicle near peoples gatherings.
- Marketing of organic products through Khadi Board outlets.
- Involvement of some NGOs like Naturopathy Centres, Yoga Centres, Rotary Clubs, Lions Clubs, Ladies Associations, Religious Growers (ISKON, Jain Communities, Sindhi Communities for publicity education)
- Development of Tourist villages.
- Formation of associations of organic producers and consumers.
- Adequate publicity and training.

Questions

1. Give a brief idea of the world organic market?
2. What are the key characteristics of organic market?
3. What is status of organic export market in India?
4. What are the limitations of domestic organic market in India?
5. Why there was increase in the consumers for organic products in the West countries?
6. Which are the countries having most of organic farming area under cultivation?
7. Which are the largest international markets for organic products in the world?
8. How distribution of organic products takes place in the market?
9. What are major organic food products available in the Indian market?
10. Fill in the blanks
    a) Area under organic farming in India is.......................
    b) Total quantity of organic product in India is..................
    c) Major export produce from plantations are....................
    d) Major spices exported from India are ....
    e) The domestic markets for organic products are...........
    f) Major agency promoting organic farming in the private sector is ....
    g) The major organic export markets are.......................
    h) Trade with organic products are showing growth rates over ....per cent which are rarely found in food markets.
    i) Organic .... and .... are the major organic products desired by Indian customers.

References

Abrol, I.P. and Palaniappan, S.P. (1987). Green manure crops in irrigated and rainfed lowland rice based cropping systems in South Asia. *In:* Green Manure in Rice Farming. Proc. of a symposium on "Sustainable Agriculture", IRRI, Los Banos, Philippines.

Acharya, C.N. (1939). Comparision of different method of composting waste materials. Indian Journal of Agriculture Soil, 9:565-572.

Alvares, Claude (1996). Genes and the Environment, in Anon, The Organic Source Book, The Other India Press, Mapusa, Goa.

Bisoyi, B.N.; Majumdar, P.; Srivastava, R.S.M. and Creep, S. (2003). Prospects of Organic Input Production in India, Presented at National Seminar on Organic Farming with Special Reference to Organic Inputs, 21- 22, August, Bangalore.

Deb, D.L. and Joshi, H.C. (1994). Environment Pollution and Agriculture. *In* Natural Resource Management for Sustainable Agriculture and Environment, Angkor Publications (P) Ltd., New Delhi

Gaur, A.C.; Neelkamal, S. and Dargan, K.S. (1995). Organic manures, ICAR Publication, New Delhi.

Hegde, B.R.; Krishnegowda, K.T. and Parvathappa, H.C. (1995). Organic Residue Management in Red Soils under Dryland Conditions. *In* Shivashankaar, K. (ed.). Alternatives to Fertilizers in Sustainable Agriculture, University of Agricultural Sciences, Bangalore.

Kathpal, T.S. and Kumari, Beena (1997). Pesticidal Contamination of Environment, Its Implications and Management, in Ecological Agriculture and Sustainable Development, Indian Ecological Society and Centre for Research in Rural Development, Chandigarh.

Naraynan, S. (2005). Organic Farming in India: Relevance, Problems and Constraints, NABARD.

Prakash, T.N. (2003). A Theoretical Framework to Promote Organic Produce Marketing in India, Indian Journal of Agriculture Marketing, Conference Number Special.

Singh, A.K. (2007). Conversion to Organic Agriculture. International Book Distributing Co. Lucknow.

Singh, A.K. (2008). Production Technology on Bio-Organic Farm Inputs. International Book Distributing Co. Lucknow.

Thakur, D.S.; Sharma K.D.; Thakur, D.R. and Saini, A.S. (2003). Economics of Production and Marketing of Organic Produce and the Burgeoning Niche Market for Organic Food Products, Indian Journal of Agriculture Marketing, Conference Number Special.

Veeresh, G.K. (1999). Organic Farming Ecologically Sound and Economically Sustainable, Plant Horti. Tech, 1(3), Nov-Dec.

Willer, H. (2008). Organic agriculture worldwide: current statistics. In: The World of Organic Agriculture. Statistics and Emerging Trends, IFOAM, Bonn and FiBL, Frick, pp. 23-46.

Appendices

(A) Permitted/Restricted Products Under Certified Organic Farming (source: www.apeda.com)

APPENDIX-1

Products for Use in Fertilising and Soil Conditioning

Matter Produced on an Organic Farm Unit	
Farmyard & poultry manure, slurry, urine	Permitted
Crop residues and green manure	Permitted
Straw and other mulches	Permitted
Matter Produced Outside the Organic Farm Unit	
Blood meal, meat meal, bone meal and feather meal without Preservatives	Restricted
Compost made from any carbon based residues (animal excrement including poultry)	Restricted
Farmyard manure, slurry, urine *(Preferably after control fermentation and /or appropriate dilution) "Factory" farming sources not permitted.)*	Restricted
Fish and fish products without preservatives	Restricted
Guano	Restricted
Human excrement	Not Allowed
By-products from the food and textile industries of biodegradable material of microbial, plant or animal origin without any synthetic additives	Restricted
Peat without synthetic additives (prohibited for soil conditioning)	
Sawdust, wood shavings, wood provided it comes from untreated wood	
Seaweed and seaweed products obtained by physical processes, extraction with water or aqueous acid and/or alkaline solution	Restricted
Sewage sludge and urban composts from separated sources which are monitored for contamination	Restricted
Straw	Restricted
Vermicasts	Restricted
Animal charcoal	Restricted
Compost and spent mushroom and vermiculate substances	Restricted

Contd...

Compost from organic household reference	Restricted
Compost from plant residues	
By products from oil palm, coconut and cocoa (including empty fruit bunch, palm oil mill effluent (pome), cocoa, peat and empty cocoa pods)	Restricted
By products of industries processing ingredients from organic agriculture	Restricted
Minerals	
Basic slag	Restricted
Calcareous and magnesium rock	Restricted
Calcified seaweed	
Calcium chloride	
Calcium carbonate of network origin (chalk, limestone, gypsum and phosphate chalk)	
Mineral potassium with low chlorine content (e.g. sulphate of potash, kainite, sylvinite, patenkali)	Restricted
Natural phosphates (e.g. Rock phosphates)	Restricted
Pulverised rock	Restricted
Sodium chloride	
Trace elements (boron, Fe, Mn, molybdenum, Zn)	Restricted
Wood ash from untreated wood	Restricted
Potassium sulphate	Restricted
Magnesium sulphate (Epson salt)	
Gypsum (calcium sulphate)	
Stillage and stillage extract	Ammonium Stillage excluded
Aluminium calcium phosphate	Restricted
Sulphur	Restricted
Stone mill	Restricted
Clay (bentonite, perlite, zeolite)	
Microbiological Prepararations	
Bacterial preparations (biofertilizers)	
Biodynamic preparations	
Plant preparations and botanical extracts	
Vermiculate	
Peat	

"Factory" farming refers to industrial management systems that are heavily reliant on veterinary and feed inputs not permitted in organic agriculture.

APPENDIX-2

Products for Plant Pest and Disease Control

Certain products are allowed for use in organic agriculture for the control of pests and diseases in plant production. Such products should only be used when absolutely necessary and should be chosen taking the environmental impact into consideration.

Many of these products are restricted for use in organic production. In this appendix **"restricted"** means that the conditions and the procedure for use shall be set by the certification programme.

I. Substances from plant and animal origin	
Preparation of rotenone from Derris elliptica, Lonchocarpus, Thephrosia spp.	Restricted
Gelatine	
Propolis	Restricted
Plant based extracts (e.g. neem, garlic, pongamia, etc.)	
Preparation on basis of pyrethrins extracted from Chrysanthemum cinerariaefolium, containing possibly a synergist pyrethrum cinerafolium	Restricted
Preparation from Quassia amara	Restricted
Release of parasite predators of insect pests	Restricted
Preparation from Ryania species	Restricted
Lecithin	Restricted
Casein	
Sea weeds, sea weed meal, sea weed extracts, Sea salt and salty water	Restricted
Extract from mushroom (Shitake fungus)	
Extract from Chlorella	
Fermented product from Aspergillus	Restricted
Natural acids (vinegar)	Restricted
II. Minerals	
Chloride of lime/soda	Restricted
Clay (e.g. bentonite, perlite, vermiculite, zeolite)	
Copper salts/inorganic salts (Bordeaux mix, copper hydroxide, copper oxychloride) used as a fungicide maximum 8 kg per ha per year depending upon the crop and under the supervision of inspection and certification agency	Restricted
Mineral powders (stone meal, silicates) Not allowed	Not allowed
Diatomaceous earth	
Light mineral oils	Restricted

Permanganate of potash	Restricted
Lime sulphur (calcium polysulphide	Restricted
Silicates (sodium silicate, quartz)	Restricted
Sodium bicarbonate	
Sulphur (as a fungicide, acaricide, repellent)	Restricted
III. Microorganism /Biocontrol agents	
Viral preparations (e.g., Granulosis viruses, Nuclear polyhydrosis, viruses etc.). • Fungal preparations (e.g., Trichoderma spp. etc.) • Bacterial preparations (e.g., Bacillus spp etc.) • Parasites, predators and sterilized insects."	
IV. Others	
Carbon dioxide and nitrogen gas	Restricted
Soft soap (potassium soap)	
Ethyl alcohol	Not allowed
Homeopathic and Ayurvedic preparations	
Herbal and biodynamic preparations	
V. Traps	
Physical methods (e.g., chromatic traps, mechanical traps, light traps, sticky traps)	
Mulches, nets	

APPENDIX-3

Criteria for the Evaluation of Additional Inputs to Organic Agriculture

Appendices 1 & 2 refer to products for fertilising of the soil and control of plant pest and diseases in organic agriculture. But there may well be other products which may be useful and appropriate for use in organic agriculture which may not fall under these headings. Appendix 3 outlines the procedure to evaluate other inputs into organic production.

The following checklist should be used for amending the permitted substance list for fertilising the soil conditioning purposes:

- The material is essential for achieving or maintaining soil fertility or to fulfil specific nutrient requirements, for specific soil-conditioning and rotation purposes which cannot be satisfied by the practices outlined in the standard or of other products included in Appendix 1. The ingredients are of plant, animal, microbial or mineral origin which may undergo the following processes:
- Physical (mechanical, thermal)
- Enzymatic
- Microbial (composting, digestion)

  and,
- Their use does not result in, or contribute to, unacceptable effects on, or contamination of, the environment, including soil organisms.

  Their use has no unacceptable effect on the quality and safety of the final product.

  The following checklist should be used for amending the permitted substance list for the purpose of plant disease or pest and weed control:
- The material is essential for the control of a harmful organism or a particular disease for which other biological, physical or plant breeding alternatives and/or effective management techniques are not available

  and,
- The substances (active compound) should be plant, animal, microbial or mineral origin which may undergo the following processes:
  - physical
  - enzymatic
  - microbial

and
- their use does not result in, or contribute to, unacceptable effects on, or contamination of, the environment.

- Nature identical products such as pheromones, which are chemically synthesised, may be considered if the products are not available in sufficient quantities in their natural farm, provided that the conditions for their use do not directly or indirectly contribute to contamination of the environment or the product.

Evaluation

When an input is to be evaluated it must first be investigated by certification programmes to see whether it fulfils the following six criteria. An input must fulfil all 6 requirements before it can be accepted as suitable for use in organic agriculture.

Inputs should be evaluated regularly and weighed against alternatives. This process of regular evaluation should result in organic production becoming ever more friendly to humans, animals, environment and the ecosystem.

1. Necessity

The necessity of each input must be established. This will be investigated in the context in which the product will be used. Arguments to prove the necessity of an input may be drawn from such criteria as yield, product quality, environmental safety, ecological protection, landscape, human and animal welfare.

The use of an input may be restricted to:

- Specific crops (especially perennial crops)
- Specific regions
- Specific conditions under which the input may be used

2. Natures and Method of Production

Nature: The origin of the input should usually be (in order of preference):

- Organic - vegetative, animal, microbial
- Mineral

Non-natural products which are chemically synthesised and identical to natural products may be used.

When there is any choice, renewable inputs are preferred. The next best choice is inputs of mineral origin and the third choice is inputs which are chemically identical to natural products. There may be ecological, technical or economic arguments to take into consideration in the allowance of chemically identical inputs.

Method of Production: The ingredients of the inputs may undergo the following processes:

- Mechanical
- Physical
- Enzymatic
- Action of micro-organisms
- Chemical (as an exception and restricted)

Collection: The collection of the raw materials comprising the input must not affect the stability of the natural habitat nor affect the maintenance of any species within the collection area.

3. Environment

Environmental Safety: The input must not be harmful or have a lasting negative impact on the environment. Nor should the input give rise to unacceptable pollution of surface or ground water, air or soil. All stages during processing, use and breakdown must be evaluated.

The following characteristics of the input must be taken into account:

Degradability: All inputs must be degradable to their mineral form. Inputs with a high acute toxicity to non-target organisms should have a maximum half-life of five days. Natural substances used as inputs which are not considered toxic do not need to be degradable within a limited time.

Acute toxicity to non-target organisms: When inputs have a relatively high acute toxicity for non-target organisms, a restriction for their use is needed. Measures have to be taken to guarantee the survival of these non-target organisms. Maximum amounts allowed for application may be set. When it is not possible to take adequate measures, the use of the input must not be allowed.

Long-term chronic toxicity: Inputs which accumulate in organisms or systems of organisms and inputs which have, or are suspected of having, mutagenic or carcinogenic properties must not be used. If there are any risks, sufficient measures have to be taken to reduce any risk to an acceptable level and to prevent long lasting negative environmental effects.

Chemically synthesised products and heavy metals: Inputs should not contain harmful amounts of man made chemicals (xenobiotic products). Chemically synthesised products may be accepted only if identical to the natural product.

Mineral inputs should contain as few heavy metals as possible. Due to the lack of any alternative, and long-standing, traditional use in organic agriculture, copper and copper salts are an exception for the time being. The use of copper in any form in organic agriculture must be seen, however, as temporary and use must be restricted with regard to environmental impact.

4. Human Health and Quality

Human Health: Inputs must not be harmful to human health. All stages during processing, use and degradation must be taken into account. Measures must be taken to reduce any risks and standards set for inputs used in organic production.

Product quality: Inputs must not have negative effects on the quality of the product - e.g. taste, appearance and quality.

5. Ethical Aspects - Animal Welfare

Inputs must not have a negative influence on the natural behaviour or physical functioning of animals kept at the farm.

6. Socio Economic Aspects

Consumers' perception: Inputs should not meet resistance or opposition of consumers of organic products. An input might be considered by consumers to be unsafe to the environment or human health, although this has not been scientifically proven. Inputs should not interfere with a general feeling or opinion about what is natural or organic - e.g. genetic engineering.

APPENDIX-4

List of Approved Ingredients of Non Agricultural Origin and Processing Aids Used in Food Processing

Food Additives and Carriers 1

Intl Numbering System	Additive/ Processing aid	Used as Additive	Used as Proc. Aid	Food category	Functions	Limitation /Note
INS 170	Calcium carbonate	*	*	GA	Anticaking, acidity regulator, emulsifier, stabiliser	
INS 220	Sulphur dioxide	*	*	W	Preservative, stabiliser	Max. 0.3mg/l
INS 224	Potassium metabisulphite	*	*	W		
INS 270	Lactic acid	*	*	FV, W, MP	Acidity regulator	
INS 290	Carbon-dioxide	*	*	GA	Carbonating agent, packing gas	
INS 296	Malic acid	*	-	FV, MP	Acidulent	
INS 300	Ascorbic acid	*	-	GA	Antioxidant,	
INS 306	Tocopheroles, mixed natural concentrates	*	-	GA	Antioxidant	
INS 322	Lecithin	*	*	GA	Antioxidant, emulsifier, stabiliser	
INS 330	Citric acid	*	-	FV, MP W	Acidity regulator Antioxidant	 Restricted 1 gm/l
INS 335	Sodium citrate	*	-	ME, MP	Acidity regulator Antioxidant	
INS 336	Potassium citrate	*	-	ME, MP	Acidity regulator Antioxidant	
INS 400	Aliginic acid	*	-	FV	Emulsifier, Stabiliser, Thickener	

1 Food Additives may contain carriers which shall be evaluated

INS 401	Sodium alginate	*	-			
INS 402	Potassium alginate	*	-			
INS 333	Calcium citrate	*	-	ME, MP	Acidity regulator Antioxidant	
INS 334	Tartaric acid	*	*	W, MP	Flour treatment, raising agent, emulsifier, antioxidant, preservative	
INS 407 INS 335	Sodium tartrate	*	*	CO/CB		
INS 336	Potassium tartrate	*	*	C/CO/ CB		
INS 341	Mono calcium phosphate	*	-	C	Only for raising flour	
INS 342	Ammonium phosphate	*	-	W	-	0.3 gm/l
INS 406	Agar	*	-	MP, F	Emulsifier, Stabiliser Thickner	Max. 0.5%
INS407	Carrageenan	*	-	MP, F		
INS 410	Locust bean gum	*	-	MP, F		
INS 412	Guar gum	*	-	MP, F		
INS 413	Tragacanth gum	*	-	GA	Emulsifier, Stabiliser Thickner	Max. 0.5%
INS 414	Arabic gum	*	-	MP, F	Emulsifier, Stabiliser Thickner	Max. 0.5%
INS 415	Xanthan gum	*	-	F/FV/ CB	Emulsifier, Stabiliser Thickner	Max. 0.5%
INS 416	Karaya gum	*	-	MP, F		
INS 440	Pectin (Unmodified form)	*	-	FV, F, CB	Emulsifier, Stabiliser, Thickner	
INS 500	Sodium carbonate	*	*	CO/CB	Acidity regulator, Stabiliser, Anticaking and Raising agent	
INS 501	Potassium carbonate	*	*	C/CO/ CB	Acidity regulator Stabiliser for drying grapes	
INS 503	Ammonium carbonate	*	-	C/CO/ CB	Acidity regulator Stabiliser, Raising agent	

INS 504	Magnesium carbonate	*	-	C/CO/CB	Acidity regulator Stabiliser	
INS 508	Potassium chloride	*	-	FV	Stabiliser, Thickner	
INS 509	Calcium chloride	*	-	ME/F/ FV/SO	Coagulation, firming agent	
INS 516	Calcium sulphate	*	*	CB, SO	Acidity regulator, Stabiliser , Flour treatment, coagulation agent.	Restricted Only in baker's' yeast
INS 517	Ammonium sulphate	*		W	Encourage the growth of yeasts	Restricted to 0.3 gm/1
INS 524	Sodium hydroxide	*	*	C, S, Flours & Starches	Acidity regulator, Surface treatment of traditional bakery products. Processing aid for sugar	
INS 526	Calcium hydroxide	*	*	S, C	Additive for maize tortilla flour. Processing aid for sugar	
INS 938	Argon	*	-	GA		
INS 941	Nitrogen	*	*	GA		
INS 948	Oxygen	*	*	GA		
INS153	Wood ash	-	*	MP	Coating agent	
INS 181	Tannin	_	*	W	Clarifying agent	
INS 184	Tannic Acid	-	*	W	Filtration aid	
INS 513	Sulphuric acid	-	*	S	pH adjustment of water in sugar production.	
INS 551	Silicon dioxide	-	*	W/ Dehydra-ted FV Herbs & Spices	Gel or colloidal solution Anticaking agent	
INS 553	Talc	-	*	GA	Lubricant	0.5% -PFA
INS 901	Beeswax	-	*	GA	Releasing agent	

INS 903	Carnauba wax	-	*	GA	Releasing agent	
	Activated carbon	-	*	GA	Decolorizer	
	Bentonite	-	*	FV/W	Filter aid, clarifying agent	
	Casein	-	*	W	Clarifying agent	
	Diatomaceous earth	-	*	S/FV/W	Filter aid	
	Egg white albumen	-	*	W	Clarifying agent	
	Ethanol	-	*	GA	Solvent	
	Gelatine	-	*	MP, FV/W	Emulsifier, Clarifying agent	
	Isinglass	-	*	W	Clarifying agent	
	Kaolin	-	*	GA	Filter aid, Extraction of propolis	
	Perlite	-	*	GA	Clarifying agent	
	Preparations of bark	-	*	S		
	Vegetable oils	-	*	GA	Greasing, releasing agent	
	Glycerol	-	*		Plant extracts Restricted	
	Beet sugar	-	*		Restricted	
	Natural Colours					
	Carotenoids					
	Chlorophyll					
	Annatto					
	Saffron					
	Riboflavin					
	(Lactoflavin)					

	Curcumin Caramel Canthaxanthin					

Key - list of abbreviations used in above tables:

* Could be used as:

\- Not used as:

GA - Generally Unrestricted	MP - Milk Products
F - Fat products	ME - Meat Products
C - Cereal Products	FV - Fruit/ Vegetable
W - Wine S - Sugar products	CO - Confectionery
CB - Cakes and Biscuits	SO - Soybean products

Flavouring Agents

- Volatile (essential) oils produced by means of solvents such as oil, water, ethanol, carbon dioxide and mechanical & physical processes
- Natural smoke flavour
- Use of natural flavouring preparations should be approved as per national procedure to evaluate additives and processing aids based criteria for evaluation of additional inputs to organic agriculture (Appendix-3)

Preparations of Microorganisms

- Preparations of microorganisms accepted for use in food processing. Genetically modified organisms are excluded.
- Bakers yeast produced without bleaches and organic solvents.

Preparations of Microorganisms and Enzymes

These may be used as processing aids with approval based on the national procedure to Evaluate Additives and Processing Aids for Organic Food Products.

Ingredients

- Drinking water
- Salts (with sodium chloride and potassium chloride as basic component generally used in food processing).
- Minerals (including trace elements) and vitamins, fatty acids, amino acid and other nitrogenous compounds where their use is legally required or where severe dietary or nutritional deficiency can be demonstrated.

APPENDIX-5

Criteria for the Evaluation of Additives and Processing Aids for Organic Food Products

Introduction

List of additives, processing aids, flavouring agents and colours in organic food products are listed in Appendix-4. The following aspects and criteria should be used for evaluation of additives and processing aids in organic food products.

1. Necessity

Additives and processing aids may only be allowed in organic food products if each additive or processing aid is essential to the production wherein the authenticity of the product is respected and the product cannot be produced or preserved without them.

2. Criteria for the approval of additives and processing aids.

The additives and processing aid may be used in the processing of organic foods where:

- There are no other acceptable technologies available to process or preserve the organic product.
- The use of additives or processing aids which minimize physical or mechanical damage to the foodstuff as a substitute for other technologies which if used would result in such damage.
- The hygiene of the product cannot be guaranteed as effectively by other methods (such as a reduction in distribution time or improvement of storage facilities).
- Additives or processing aids do not compromise the authenticity of the product.
- The additives or processing aids do not confuse the customer by giving the impression that the final product is of higher quality than is justified by the quality of the raw material. This refers primarily but not exclusively, to colouring and flavouring agents.
- Additives and processing aids should not detract from the overall quality of the product.
- The additives generally shall have 'GRAS' (Generally Regarded As Safe) status indicating that they are safe when used in accordance with good manufacturing practices.

3. Step by step procedure for use of additives and processing aids

a) Instead of using additives or processing aids, the preferred first choice is

- Food grown under organic conditions which are used as a whole product or are processed in accordance with the IFOAM basic standards – e.g. flour used as a thickening agent or vegetable oil as a releasing agent.
- Foods or raw materials of plant and animal origin which are produced only by mechanical or simple physical procedures – e.g. salt.

b) The second choice is:

- Substance isolated from food and produced physically or by enzymes – e.g. starch, tartrates, pectin.
- Purified products of raw materials of non-agricultural origin and microorganisms – e.g. acerola fruit extract, enzymes and microorganism preparations such as starter cultures.

c) In organic food products the following categories of additives and processing aids are not allowed:

- "Nature identical" substances
- Synthetic substances primarily judged as being unnatural or as a "new construction" of food compounds such as acetylated cross linked starches (modified starches).
- Additives or processing aids produced by means of genetic engineering
- Synthetic colouring and synthetic preservatives.
- Carriers and preservatives used in the preparation of additives and processing aids shall also be taken into consideration.

APPENDIX-6

Approved Additives for Manufacturing of Packaging Films for Packaging of Organic Foodstuffs

Certain additives are allowed for use in manufacture of packaging films for packaging of foodstuffs. However, many of these are restricted for use in packaging of organic foodstuffs. Restricted means that the conditions and procedures for use shall be set by the accredited certification programme.

The following are approved additives under restriction: -

Use of plastics for packaging of organic foodstuffs

S. No.	Products	Limitation
1.	4,4'-Bis(2-benzoxazolyl)stilbene	Restricted
2.	9,9-Bis(methoxymethyl)fluorine	Restricted
3.	Carbonic acid, copper salt	Restricted
4.	Diethyleneglycol	Restricted
5.	2-(4,6-Diphenyl-1,3,5-triazin-2-yl)-5-(hexyloxy)phenol	Restricted
6.	Ethylenediaminetetraacetic acid, copper salt	Restricted
7.	2-(2-Hydroxy-3,5-di-tert-butyl-phenyl-5-chlorobenzotriazole	Restricted
8.	2-Methyl-4-isothiazolin-3-one	Restricted
9.	Phosphoric acid, trichlorocthylester	Restricted
10.	Polyesters of 1,2 propanediol and/or 1,3-and 1, 4 butanediol and/or polypropyleneglycol with adipic acid, also end-capped with acetic acid or fatty acids C10-C18 or n-octanol and/or n-decanol	Restricted
11.	1,1,1-Trimethylolpropane	Restricted
12.	3-hydroxybutanoic acid 3-hydro xypentanoic acid, copolymer	Restricted

APPENDIX-7

List of Approved Feed Materials, Feed Additives and Processing Aids for Animal Nutrition

1.	**Feed Materials of Plant Origin**	
	• Cereals grains, their products and bi-products	
	• Oilseeds, oil fruits, their products and bi-products	
	• Legume seed, their products and bi-products	
	• Tuber roots, their products and bi-products	
	• Other seeds and fruits	
	• Forages and roughages	
	• Molasses as a binding agent	
2.	**Feed Material as an Animal Origin**	
	• Milk and milk products	
	• Fish, other marine animals, their products and bi-products	
3.	**Feed Material from Mineral Origin**	
	• Sea salt, rock salt	Restricted
	• Sodium sulphate	Restricted
	• Sodium carbonate	Restricted
	• Sodium bicarbonate	Restricted
	• Sodium chloride	Restricted
	• Calcium carbonate	Restricted
	• Calcium lactate	Restricted
	• Calcium gluconate	Restricted
	• Bone dicalcium phosphat precipitate	Restricted
	• Defluorinated dicalcium phosphate	Restricted
	• Defluorinated monocalcium phosphate	Restricted
	• Anhydrous magnesia	
	• Magnesium sulphate	Restricted
	• Magnesium chloride	Restricted
	• Magnesium carbonate	Restricted
4.	**Trace Elements**	
	• Iron	Feed additives
	• Iodine	Feed additives
	• Cobalt	Feed additives
	• Manganese	Feed additives

	• Zinc • Molybdenum • Selenium	Feed additives Feed additives Feed additives
5.	**Vitamins**	Restricted
6.	**Enzymes**	Restricted
7.	**Micro-organisms**	Restricted
8.	**Preservatives** • E-336 Formic acid • E-260 Acetic acid • E-270 Lactic acid • E-280 Propionic acid	
9.	**Binders, anti-caking agent and coagulants** • E-551b Colloidal silica • E-551c Kieselgur • E-553 Sepiolite • E-558 Bentonite • E-559 Kaolinitic clays • E-561 Vermiculite • E-599 Perlite	
10.	**Processing aids for silage** • Sea salt • Coarse rock salt • Enzymes • Yeasts • Sugar • Sugar beet pulp • Cereal flour • Molasses • Lactic	

APPENDIX-8

Products Authorized for Cleaning and Disinfections of Livestock Buildings and Installations

- Potassium and sodium soap
- Water and steam
- Milk of lime
- Lime
- Quicklime
- Sodium hypochlorite (e.g. as liquid bleach)
- Caustic potash
- Hydrogen peroxide
- Natural essences of plants
- Citric, peracetic acid, formic, lactic, oxalic and acetic acid
- Alcohol
- Nitric acid (diary equipment)
- Phosporic acid (dairy equipment)
- Formaldehyde
- Sodium carbonate

(C) Accredited Inspection and Certification Agencies

APPENDIX-9

Bioinspecta
Director, Ackerstrasse, Postfach CH-5070 Frick, Switzerland,
Branch office in India: Bioinspecta, C/o INDOCERT, Thottumugham
P.O. Aluva-683105, Ernakulam, Kerala State, India

Ecocert International (Germany)
Country Representative/Managing Director
Ecocert SA Branch Office, 54 A, Kanchan Nagar,
Nakshetrawadi Aurangabad – 431002, Maharashtra State

Indian Organic Certification Agency (INDOCERT)
Executive Director, Thottumugham P.O. Aluva-683105,
Ernakulam Kerala State, India

International Resources for Fairer Trade
Director, Sona Udyog (Industrial Estate), Unit No. 7, Parsi Pandhayat Road,
Andheri (E), Mumbai – 400 069

IMO Control Private Limited
Director, 26, 17th Main, HAL 2nd 'A', Stage, Bangalore-560 008

LACON GMBH, Germany
Dr. Heinz Joachim Kopp, M.D.
Weingarten Str. 1.5, 77654, Offenburg, Germany
Branch office in India
Mr. Bobby Issac, LACON, C/o Renewable Energy Centre, Mithradham,
Chunangaveli, Alwaye-683 105, Kerala

SGS India Pvt. Ltd.
Business Manager, M/s SGS India Pvt. Ltd., 250 Udyog Vihar
Phase – IV, Gurgaon – 122015

Skal International (Netherlands)
International Inspector, Skal Inspection and Certification Agency
No. 191, 1st Main Road, Mahalaxmi Layout, Bangalore – 560086

APOG Organic Certification Agency
1st Floor, 5th Main, 9th Cross, Jayamahal Extension, Banglore-46

OneCert Asia Agri Certification Pvt. Ltd
Agrasen Farm, Vatika Road, off. Tonk Road,(P.O.Vatika)
Jaipur - 303 905, Rajasthan, India.

(E) Useful Websites on Organic Agriculture

APPENDIX-10

1. http://www.ifoam.org (IFOAM-International Federation of Organic Agriculture Movements)
2. http://www.ioia.net (IOIA-Independent Organic Inspectors Association)
3. http://www.organicts.com/organic info/certification/links/index.html
4. http://www.organic-research.com [EU Regulation 2092/91]
5. http://www.ams.usda.gov/nop [State and Federal Organic Programs In The United States]
6. http://www.ota.com [ORGANIC Trade Association-American Organic Standards]
7. http://www.fao.org/organicag [FAO (website on organic agriculture)
8. http://home.prolink.de/~hps/organic/consolid-en.html (EU regulation can be downloaded from this website)
9. http://www.agrecol.de : http://www.agrecolandes.org/
10. http://www.agromisa.org
11. http://www.universalorganics.com.au (general information on organic certification of Australia)
12. http://www.cabi-bioscience.org
13. http://fgf@global.co.za/home
14. http://www.ofa.org.au (organic federation of Australia)
15. http://www.organicherbs.org (organic herb growers of Australia)
16. http://www.sustainablecap.dk (sustainable cap initiative in Denmark)
17. http://www.opam.mb.ca (organic producers association of Manitoba)
18. http://www.gtz.de
19. http://www.zadi.de (German Centre for Documentation and Information in Agriculture)
20. http://www.varanashi.com (varanashi research foundation, India)
21. http://www.dpw.wau.nl/biob (biological farming systems research and teaching group, Wageningen University, the Netherlands)
22. http://www.hdra.org.uk/

23. http://www.geocities.com/rainforest/vines/6274/
24. http://www.cabi.org (abstracts on organic farming in temperate regions)
25. http://www.fibl.ch (Research Institute on Organic Agriculture, Switzerland))
26. http://www.napcl.com (Nadukkara Agro Processing Co. Ltd)
27. http://www.apeda.com (Agricultural and Processed Food Products Export Development Authority, India)
28. http://www.biofach.de (World trade fair for organic foods and natural products)
29. http://www.hortibizindia.nic.in (National Horticulture Board, India)
30. http://www.kinfra.com (KINFRA Food Processing Parks)
31. http://www.iide.org/
32. http://www.kenyaweb.com/agriculture/organic-agri/
33. http://www.hivos.nl (Economic and cultural development in Africa, Asia, Latin America and Southeast Europe)
34. http://www.intercooperation.ch (Intercooperation-IC)
35. http://www.andrewlorand.com (Consulting for Ecological & Biodynamic Agriculture)
36. http://www.indianorganic.info (International conference –Indian organic products)
37. http://www.lifecyclesproject.ca (Life Cycles, Victoria)
38. http://www.ibl.ch-internat/
39. http://www.maela-net.org/
40. http://www.eda.admin.ch/newdelhi (Economic and commercial affairs, embassy of Switzerland, India)
41. http://www.oekoskolen.dk (The Organic Agriculture College, Denmark)
42. http://www.naturland.de (Naturland-organic certifier)
43. http://www.phaladaagro.com (Phalada Agro Research Foundation Ltd, India)
44. http://www.seco-admin.ch (State Secretariat for Economic Affairs, Switzerland)

45. http://www.shl.bfh.ch/

46. http://members.shaw.ca/gardenab (Organic seeds and crops researcher)

47. http://www.indianspices.com (Spices Board, India)

48. http://www.organic-farm.com (Hong Kong organic farming association)

49. http://www.unitec.ac.nz (School of Landscape & Plant Science Sustainable Organic Systems Design Integrated Studies)

50. http://www.worldcom@island.net (World Community Development Education Society)

51. http://forum.yam.org.tw (Homemaker's Union and Foundation, Taiwan)

52. http://www.biosuisse.com (Swiss Inspection and Certification Body)

53. http://www.bio-inspecta.ch/ (The Swiss Organic Certification body

54. http://www.vso.org.uk/

55. http://www.wn.org/

56. http://www.apeda.com/ (Indian Organic Standard NPOP-English and Hindi, can be downloaded from this website)

FSC
www.fsc.org
MIX
Papier | Fördert
gute Waldnutzung
FSC® C083411

Zeitfracht Medien GmbH
Ferdinand-Jühlke-Straße 7
99095 Erfurt, Deutschland
produktsicherheit@kolibri360.de